NEW PROBLEMS IN ASTROMETRY

INTERNATIONAL ASTRONOMICAL UNION
UNION ASTRONOMIQUE INTERNATIONALE

SYMPOSIUM No. 61

HELD IN PERTH, WESTERN AUSTRALIA, 13–17 AUGUST, 1973

NEW PROBLEMS IN ASTROMETRY

EDITED BY

WILHELM GLIESE

Astronomisches Rechen-Institut, Heidelberg, F.R. Germany

C. ANDREW MURRAY

Royal Greenwich Observatory, Herstmonceux, England

R. H. TUCKER

Royal Greenwich Observatory, Herstmonceux, England

D. REIDEL PUBLISHING COMPANY

DORDRECHT-HOLLAND / BOSTON-U.S.A.

1974

Published on behalf of
the International Astronomical Union
by
D. Reidel Publishing Company, P.O. Box 17, Dordrecht, Holland

Sold and distributed in the U.S.A., Canada, and Mexico
by D. Reidel Publishing Company, Inc.
306 Dartmouth Street, Boston,
Mass. 02116, U.S.A.

Library of Congress Catalog Card Number 73–94453

Cloth edition: ISBN 90 277 0444 9
Paperback edition: ISBN 90 277 0445 7

Printed in The Netherlands by D. Reidel, Dordrecht

TABLE OF CONTENTS

SESSION C / RADIO ASTROMETRY

SESSION D / ASTROMETRY WITH LARGE TELESCOPES

GENERAL DISCUSSION, SUMMARY AND RESOLUTIONS

PREFACE

The IAU Symposium No. 61, *New Problems in Astrometry*, was held in Perth, Western Australia, from 13th to 17th August 1973, under the sponsorship of Commission 8 (Positional Astronomy), 24 (Photographic Astrometry), 33 (Structure and Dynamics of the Galactic System) and 40 (Radio Astronomy). Considerable financial support was given by the Government of Western Australia. The scientific organising committee included Prof. Walter Fricke (Chairman), Prof. S. Vasilevskis (Vice-Chairman), Dr W. Gliese and Mr C. A. Murray (Secretaries), Dr B. G. Clark, Mr B. Elsmore, Prof. S. W. McCuskey, Dr J. L. Schombert, Prof. R. H. Stoy, Dr K. Aa. Strand, Dr G. van Herk, Mr H. W. Wood and Prof. M. S. Zverev. The local organising committee included Mr B. J. Harris (Chairman), Mr P. V. Birch, Mr M. P. Candy, Mr D. N. Harwood, Dr I. Nikoloff and Dr S. E. Williams. The duties of chairmen of the sessions were performed by: B. L. Klock, W. Fricke, H. W. Wood, C. A. Anguita, C. M. Wade, B. Elsmore, B. J. Harris, J. Kovalevsky, S. Vasilevskis, G. van Herk, K. Aa. Strand and C. A. Murray.

This volume contains the texts or abstracts of papers presented at the Symposium. These reflect not only the continuing interest in the classical methods of optical astrometry, but also the exciting developments already achieved or to be expected within the next few years through the new techniques of radio interferometry, the use of large optical telescopes and space astrometry. This was in fact the first symposium in which both optical and radio astrometrists participated. Collaboration between radio and optical observers, and the exploitation of these new techniques, were urged in three general Resolutions which were adopted at the end of the Symposium.

Unfortunately several of those invited, including some members of the scientific organising committee, were prevented from attending the Symposium; nevertheless papers or abstracts which they submitted have been included in this volume. Among these are several papers from our Russian colleagues, none of whom was able to be present.

A special debt of gratitude is owed by all the participants to Mr Harris and his colleagues for the efficient organisation of the Symposium, and, as Secretaries, we are particularly grateful for the assistance given to us in recording the discussions which appear in this volume. We also wish to acknowledge with thanks the valuable help given by Mrs M. Laker in the preparation of the manuscripts for publication.

W. GLIESE
C. A. MURRAY
R. H. TUCKER

LIST OF PARTICIPANTS

Anguita, C. A.	Santiago, Chile
Baars, J. W. M.	Westerbork, The Netherlands
Birch, P. V.	Perth, W. Australia
Blanco, V. M.	Cerro Tololo, Chile
Bok, B. J.	Tucson, Arizona, U.S.A.
Bok, P. F.	Tucson, Arizona, U.S.A.
Brosche, P.	Bonn, F.R. Germany
Brouw, W. N.	Canberra, A.C.T., Australia
Candy, M. P.	Perth, W. Australia
Carrasco, G.	Santiago, Chile
Clube, S. V. M.	Edinburgh, Scotland
Counselman, C. C., III	Cambridge, Mass., U.S.A.
Débarbat, S.	Paris, France
Dieckvoss, W.	Hamburg, F.R. Germany
Douglas, J. N.	Austin, Texas, U.S.A.
Edmondson, F. K.	Bloomington, Indiana, U.S.A.
Eichhorn-von Wurmb, H. K.	Tampa, Florida, U.S.A.
Elsmore, B.	Cambridge, England
Fracastoro, M. G.	Turin, Italy
Franz, O. G.	Flagstaff, Arizona, U.S.A.
Fricke, W.	Heidelberg, F.R. Germany
Gliese, W.	Heidelberg, F.R. Germany
Gubbay, J.	Salisbury, S. Australia
Harris, B. J.	Perth, W. Australia
Harwood, D. N.	Perth, W. Australia
Heintz, W. D.	Swarthmore, Pennsylvania, U.S.A.
Hoffleit, E. D.	New Haven, Connecticut, U.S.A.
Hunstead, R. W.	Sydney, N.S.W., Australia
Klemola, A. R.	Santa Cruz, California, U.S.A.
Klock, B. L.	Washington, D.C., U.S.A.
Kovalevsky, J.	Meudon, France
Lacroute, P.	Strasbourg, France
Lederle, T.	Heidelberg, F.R. Germany
Legg, A. J.	Salisbury, S. Australia
Luck, J. McK.	Canberra, A.C.T., Australia
Luyten, W. J.	Minneapolis, Minnesota, U.S.A.
Matsunami, N.	Tokyo, Japan

Moffet, A. T.	Pasadena, California, U.S.A.
Moutsoulas, M.	Athens, Greece
Mulholland, J. D.	Austin, Texas, U.S.A.
Muller, P.	Meudon, France
Murray, C. A.	Herstmonceux, England
Nikoloff, I.	Perth, W. Australia
Oja, T.	Kvistaberg, Sweden
Palmer, H. P.	Charlottesville, Virginia, U.S.A.
Prochazka, F.	Vienna, Austria
Robertson, D. S.	Salisbury, S. Australia
Sarma, M. B. K.	Hyderabad, India
Sinzi, A. M.	Tokyo, Japan
Stoy, R. H.	Edinburgh, Scotland
Strand, K. Aa.	Washington, D.C., U.S.A.
Sugawa, C.	Mizusawa, Japan
Teleki, G.	Belgrade, Yugoslavia
Tucker, R. H.	Herstmonceux, England
van Altena, W. F.	Williams Bay, Wisconsin, U.S.A.
van Herk, G.	Leiden, The Netherlands
Vasilevskis, S.	Santa Cruz, California, U.S.A.
Vicente, R. O.	Algés, Portugal
Wade, C. M.	Charlottesville, Virginia, U.S.A.
Wall, J. V.	Epping, N.S.W., Australia
Walter, H. G.	Darmstadt, F.R. Germany
West, R. M.	Geneva, Switzerland
Williams, J. G.	Pasadena, California, U.S.A.
Williams, S. E.	Perth, W. Australia
Wood, H. W.	Sydney, N.S.W., Australia
Yasuda, H.	Tokyo, Japan
Yumi, S.	Mizusawa, Japan

NEW IMPETUS TO ASTROMETRY

(Inaugural Address)

WALTER FRICKE

Astronomisches Rechen-Institut, Heidelberg, F.R. Germany

In 1964 a conference was held at Flagstaff, Arizona, in honour of Ejnar Hertzsprung. At this conference Martin Schwarzschild presented the inaugural address. The title was 'New Impetus to Astrometry'. I was so much impressed by everything what he, as an astrophysicist, had to say that I have always remembered his analysis of important astrometric problems with admiration. Nine years after the Flagstaff Conference, I find it appropriate to open this Symposium in referring to Schwarzschild's lecture and in presenting my views on recent developments and on progress which I expect in the future.

Major achievements in astrometry during the past decade resulted mainly from the introduction and development of new instrumental techniques, the application of astrometric methods in new areas of astronomy, and from the vigorous and successful work in large international undertakings. An important part of the recent progress was unforeseen, namely, the progress made in the absolute measurement of positions of objects by means of radio techniques.

Let us first consider the classical areas of astrometry. I will follow here Schwarzschild in first citing the determination of the local inertial frame of rest, which in practice, is given by a system of positions and proper motions of fundamental stars and the knowledge of accurate values of the so-called constants of precession, nutation and aberration. While in 1963 the Fourth Fundamental Catalogue (FK4) was completed on the basis of observations before about 1950, an enormous effort began with the aim of improving the system and the individual data of the FK4. The observations have been carried out with the transit circle, the vertical circle, and with Danjon's astrolabe. By means of improved techniques the errors of observations have been decreased by a factor of about two compared with the observations on which the FK4 was based, and it has become possible to push the magnitude limit of meridian circles to stars fainter than visual magnitude nine. Furthermore, considerable contributions have been made to the improvement of the system in the southern sky, which in the FK4 was based almost entirely on observations made at the Cape Observatory alone. The number of absolute and differential observations made since 1950 is already sufficiently large to justify an extensive analysis with the aim of an improvement of the FK4 and its extension from magnitude 7.5 to 9.0. Progress has also been made in determining precession. It has become clear that Newcomb's planetary precession requires a small correction due to recent improvements of the values of the planetary masses. The point has been reached where further corrections to the planetary masses cannot yield a significant change in planetary precession for all possible purposes within the next few decades. The situation is different for the lunisolar precession where we know that Newcomb's determination is affected by large errors. Two methods of determination have recently been applied, the traditional one based on fundamental proper motions which are affected by erroneous precession. and a new one based on the comparison of fundamental proper motions with those measured with respect to galaxies. Moreover, it has become evident that recently derived values of lunisolar precession differ from Newcomb's value mainly due to the differences between the declination system of the FK4 and that used by Newcomb which was essentially the declination system of the 'FK1' compiled in 1879 (*Fundamental Catalog für die Zonenbeobachtungen am Nördlichen Himmel*, abbreviated FC). From all findings, which shall not be reviewed here, I would consider it desirable that the next fundamental catalogue, the FK5, should be based on revised values of precession.

The second area of astrometry I am now going to consider is the determination of proper motions of a large number of stars for purposes of stellar kinematics. The classical approach consisting in the repetition of plates taken with astrographs has resulted in the AGK3 giving positions for the epochs 1930 and 1958 and proper motions of about 180 000 stars in the northern sky. Here is an example of work which would have hardly been possible without international cooperation, that has proved to be a strong source of impetus to major achievements. Many observatories distributed over the whole world have shared in the task of determining the positions of reference stars in the system of the FK4 by means of transit circle observations. As a result the positions of the 20 000 reference stars forming the AGK3R catalogue belong to the most accurately known at an epoch around 1960. In the AGK3, which is in print now, the average mean error of a proper motion is $\pm 0\rlap{.}''80$ per century. For the improvement of the individual proper motions in the system of the FK4 older catalogues may be used which have not had the support of a strong system of reference stars. Among such older sources are the AGK1, Astrographic, and Yale Zone catalogues whose full exploitation is a task for the near future. I am of the opinion that, in addition, a full coverage of the northern sky by overlapping plates down to photographic magnitude twelve should be made as soon as possible. If this would be done soon, the AGK3R stars can be used again as reference stars. The repeated use of the AGK3R stars requires the knowledge of their proper motions, and the larger the epoch interval counted from 1960 will become, the more accurate the proper motions of the reference stars must be. Proper motions may be derived for many of them, by using positions known from older meridian observations that were carried out in a well defined system and can be expected to be free of serious errors depending on the magnitude.

The improvement of the AGK3 proper motions by both methods, the exploitation of old catalogues and the repetition of the plates, is a necessity for stellar kinematics. The present errors of $\pm 0\rlap{.}''80$ in each coordinate which are about four times the errors in FK4, yield satisfactory tangential velocities only for stars nearer than 300 pc

$(\varepsilon_{v_t} = \pm 11.4$ km s^{-1} at $r = 300$ pc). In practice, one can expect to shift this limit to about 600 pc by the methods just mentioned.

In the southern sky, the situation with respect to a photographic reference system of positions and proper motions is not as favorable as north of the equator. The Southern Reference Star program will supply the reference star positions for plates that are being taken at the Cape and Sydney observatories. In order to provide a useful reference system in the south, the SRS program can be considered as one step only; the observations of the program stars must be continued in future with transit circles. They are not only required for providing the reference frame for photographic plates but also for the improvement of the fundamental system.

In view of the great effort required for measuring proper motions of stars with respect to the fundamental system, all attempts must be welcomed that use galaxies as representatives of an inertial frame of reference. The first results presented during the past few years from plate pairs taken at the Lick Observatory and at Pulkovo are very encouraging for the following reasons: they have allowed a comparison with the fundamental reference system via the AGK3; they contain stars much fainter than AGK3 stars in large areas over the sky accessible to these observatories; and last not least, the independent approach made at these two observatories will certainly help to get a better judgement of the reliability of the measurements. The comparison of the results obtained independently at Lick and Pulkovo indicates that the proper motions in declination are in fair agreement with each other and with the fundamental system. The proper motions in right ascension measured at Lick and at Pulkovo, however, differ from each other by more than one second of arc per century, and the mean of both is very near to the fundamental μ_α. This discrepancy is not yet explained. Furthermore, the high expectations set in the measurements with respect to galaxies call for a continuous effort in the improvement of the individual accuracy by further observations, since, at present, peculiar motions of stars with respect to galaxies are satisfactory only, if the stars are not farther away than about 300 pc.

I wish to proceed now to a third area which has been introduced into astrometry only recently, but with most exciting success. This is the area of radio astrometry, and within this area I want to draw particular attention to the new possibilities of establishing a fundamental system of positions by absolute determinations of declinations and right ascensions of compact radio sources by means of radio interferometers. Nothing else than the zero point in right ascensions has to be adopted from other sources, if one wants to adopt the vernal equinox as the zero point. Progress in radio astrometry appears to be so rapid that any statement made by a non-expert in the field, like myself, can hardly be up-to-date. In my article on 'Fundamental Systems of Positions and Proper Motions' published last year in *Ann. Rev. Astron Astrophys.* I wrote that the methods of determining absolute radio positions are near the point where they can compete with the methods of traditional fundamental astrometry. From recent information by radio astrometrists I find that this point has already been reached and that the radio methods have been developed to such a perfection that an even higher accuracy can be reached than in optical absolute measurements. We all are certainly very much looking forward to the reports of our colleagues from the field of radio astrometry in this Symposium. To my knowledge it is the first time that optical and radio astrometrists are meeting in an IAU Symposium for a discussion of joint problems, and here is a challenging opportunity of making benefit for future developments.

The new impetus offered by radio astrometry to optical measurements has widely opened the use of large telescopes for the purpose of determining accurate positions of optical counterparts of compact radio sources. Here is the opportunity of linking the astrometric system based on radio observations with the fundamental system with the aim of establishing an extragalactic inertial system represented by bright stars and optically faint objects. All what I have said about the desirability of an improvement of the FK4 and the establishment of accurate photographic reference systems over the whole sky are necessary steps in this direction. The possibility that the compact radio sources may not all be extragalactic, that their positions in the sky may not be strictly identical with those of the optical counterparts, and that some of them may have large transverse motions should not discourage us from vigorously pursuing the aim just mentioned.

There are other areas in astrometry in which the use of large telescopes has given vital impetus. These are the determination of parallaxes of faint stars, the observation of double stars, and the search for high proper motion stars. In these fields are most fascinating contributions to astrophysics by finding absolute faint stars thus providing us with data on very faint red dwarfs, on the end of the white dwarf sequence, and on very small stellar masses. I am sure that it is no overstatement, if I remind our younger colleagues that the astrometrists have one of the keys in their hands for progress in astrophysics. This holds for observers as well as for all those who are engaged in improving the techniques of observation, of measuring, and of analysis.

I am coming to the end in addressing in particular our younger colleagues. There are before us many fascinating tasks whose accomplishment is fascilitated by new and automized observing instruments, measuring machines and fast computers. You are encouraged in many ways to get in contact with colleagues from all over the world and to take part in international cooperation. There are challenging opportunities of making a major contribution to astronomical research. I would, however, be seriously misunderstood, if one would assume that these magnificent opportunities are accessible in an easy walk. The qualification for successful research can only be reached by hard work. The requirements for qualification have not diminished, in the contrary, they have grown considerably. The astrometrist is not an operator at an instrument, a measuring machine or a computer who is expecting that new impetus will come from outside. He has to govern large areas of astronomy such that he himself can give impetus to new achievements. It is my sincere hope that this Symposium will contribute to raise some enthusiasm for great accomplishments in astrometry in the future.

SESSION A

REFERENCE SYSTEMS

REPORT ON IAU COLLOQUIUM No. 20,
'MERIDIAN ASTRONOMY'

(Invited Paper)

G. VAN HERK

Leiden Observatory, The Netherlands

At the invitation of the Copenhagen University Observatory, 32 workers in the field of meridian astronomy participated in the sessions.

The procedures followed the same scheme as is used in the reports of Commission 8 of the IAU. The subjects discussed were, first, instrumentation; second, reference systems; third, international reference star programmes and special programmes; fourth, investigations.

Invited speakers were respectively B. Strömgren, W. Fricke, J. L. Schombert (the other invited speaker on international reference star programmes, M. S. Zverev, could not participate), and H. Yasuda. None of the invited Soviet Union astrometrists attended the colloquium, so the meeting lacked much valuable advice.

Basically, the topics fell into three time-correlated groups, one dealing with future work, one with the present-day situation, and one with past work.

1. Instrumentation

The future in instrumentation was discussed mainly by Strömgren, Klock and Høg.

STRØMGREN pointed out that astrometry is still important for many astrophysical questions, and hence improvement of astrometrical techniques will help the rest of astronomy. With respect to impersonal methods, he recalled Trumpler's photographic registration of transits, already known in the 20's. Soon photo-electric methods were tried, with intrinsic difficulties, such as the variation in apparent brightness of the star due to scintillation effects and the short time available to do the registrations. These difficulties were greatly overcome by the introduction of the 'flipping technique' in 1933: while the star is passing over a grid in the focal plane, the incoming light is transferred or 'flipped' at regular, registered time-intervals from one photocell to another, enabling the integration to last close to one minute of time. However, when in 1945 a choice had to be made between different techniques for the newly conceived instrument for the Copenhagen Observatory, the photographic method was still chosen, because the modern equipments of the present day were not available then. Modern counting techniques of photon events combined with computerized evaluation of the data is, today, the best possible choice. He recalled the work of Pavlov and Requième in the 60's, and the recent work by Høg. The photographic meridian circle, as it stands today – the work of Laustsen is recalled – compares favourably with the ordinary m.c.: the limiting magnitude to be reached

Gliese, Murray, and Tucker, 'New Problems in Astrometry', 7–21. All Rights Reserved.
Copyright © 1974 by the IAU.

is certainly two magnitudes fainter, and the accuracy is the same in α as in δ, and is as good or better than that obtained with the classical instrument. Moreover, no magnitude term seems to be present. It was recalled what high accuracy is needed for galactic research: freedom of systematic errors in proper motions of 0″.02 per century is looked for. (Oort's talk of 1952 was mentioned several times in this connection.) The radio astrometric techniques which have come into use lately can give absolute determination of declinations down to the 17th mag., and we need to look for possible ties between the reference system based today on the bright stars and the results obtained on faint objects. The meridian circles should be able to give positions without systematic errors down to the 11th mag. With the use of photo-electric methods we should come to the 13th or 14th mag. Making full use of the photon-events, with a band width of 400 Å, a photo-multiplier with 10% quantum efficiency, a star with $13 < m < 14$ mag. will register 40 primary photon-events per second, and with an observation time of some 30 s, we would have to evaluate an information based on 1000 or more photon-events (an opening of 20 cm is considered). If we reduce with a computer the information stored in time intervals of the order of 0ˢ.01, and choose a width of 6″ for the occulting bars in the focal-plane grid to allow for a seeing effect of 3″, the required accuracy can be reached, and we can have good positions down to 14th mag. Some 2000 quasars will be available with optical counterparts to the 17th mag. It should not be difficult to pick the meridian circle reference stars of $m \simeq 13$ close to these quasars, and now the magnitude gap is small enough to make the connection successfully. Though there is a gap of avoidance for the quasars, their number is deemed sufficient to evaluate systematic errors of the present reference system in fields 1 h wide and 10–15 deg in δ. The observation of asteroids, pursued to fainter magnitudes, will be done with higher accuracy than today, and much further away from opposition.

Visual observations are still very good when the data are offered to the eye in the correct way as Danjon's astrolabe proves.

In conclusion, absolute observations are very important, and the ultimate goal of meridian astronomy is to tie the results of these absolute observations on the bright stars at the one end to the positions of the fainter stars at the other end.

In the discussion, Tucker suggested the use of one photocell in the flipping technique, but the feared difficulties are not considered of deciding importance, and Høg tells that this flipping technique has inspired him to build his multislit micrometer. Fricke points at the gain in accuracy since 1930: then of the order of ±0″.4, now in Brorfelde: ±0″.2. Radio-astrometrists obtain today an accuracy of ±0″.6 (Wade, 1970), and it will be a big step to come down to 0″.02. The risk that the positions from optical methods and radio methods might not be the same was not considered to be serious by Strömgren as long as the radio source had a diameter less than 1″. Gliese wanted to know whether the introduction of new, heavy instrumentation to the telescope will not change the results in a systematic way. Klock was of the opinion that this will not be the case if one is careful.

KLOCK described the newly designed meridian circle of the U.S. Naval Observatory, which was at the time of the colloquium still in the hands of the manufacturers, Farrand Optical Co. Inc. The instrument was built with the most modern technical equipment and from the latest type of material, thus ensuring great stability against thermal influences. It is a 10-in. catadioptric instrument, installed on urethane-covered concrete piers. The reading of the angular position of the instrument will be done by two inductosyn-systems, a check for which is provided by a gold inlay circle read by the new automatic circle scanning system. This scanner scans in six places simultaneously. It produces digital data to 1/5000 part of a division interval ($0°05$). The repeatability is about three times as good as found for the automatic film measuring engines of the USNO, and this confirms the findings at the Copenhagen and Hamburg observatories. The scanning micrometers consist of a geared synchronous motor turning a precision screw which moves a small photodiode-slit assembly. The circle is illuminated by a pulsed Xenon-arc lamp, synchronous with the micrometer screw rotation. The electrical output corresponding to each light pulse is amplified to a level suitable for cable transmission to the control rack. The data points corresponding to the intensity levels are read into a computer each 10 ms. About 600 intensity values define a profile of three division lines on the circle and of the two reference lines attached to each scanner.

The inductosyn-system (electronic angular position transducer) yields angular information through measurement of changing induction between two plates, and is capable of a repeatability of about $0''04$. This system has been used so far as a test on the telescope movement during observing. A temperature instability is still under investigation.

In the new instrument a self-check-system is built in to test the changes in the direction of the optical axis; the flexure can thus be determined in all directions of the telescope.

Høg asks whether this system provides a real check on the connection between optical axis and circle. Klock answers this is not necessary for the purpose if the flexure is small.

HØG describes a suggested new type of horizontal meridian circle made of glass with a zero-expansion temperature coefficient. His underlying idea is to create an instrument with a residual flexure 10 times smaller than is found in the conventional meridian circle. A consequence of this attempt will be that another type of error, namely the determination of the refraction, will be improved upon. The optical parts are mirrors (a model was exhibited). The rotating mirror, covering the meridian is cemented to another flat to check the tilt of the first mentioned one and to the glass declination circle. The image forming mirror whose tilt is checked through a fourth mirror cemented to it, is held in position through rods of non-expanding metal. A minimum number of degrees of freedom is realized in the mechanical parts. The observations are carried out photo-electrically, together with the automatic auto-collimation tests. These tests should allow a sufficient accuracy even when only a short time interval is available. Studies for the best possible ways of shielding the instrument and preventing stratification of the air, are under way.

Anguita refers to the instabilities experienced in zero-expansion glass due to the slow crystallization and he wonders how well one can trust an instrument made of this material. Høg points out that the mirror in his instrument is much smaller than the large Chilean telescope Anguita is referring to, and the coefficent of flexure is 10 times smaller than for the tube, so he believes there will be a good chance the difficulties will not be too serious. Høg expects the reproducibility of the position of the micrometer to be as good as $0''.2$, or $0''.02$. The tilt of the mirror which covers the meridian should be reproducible to $2''$ and therefore this quantity must be measured after every observation of a star to an accuracy of $\pm 0''.05$ in as little as 5 seconds of time. Van Herk remarked that in an ordinary meridian circle the designer also hopes to have zero flexure; it is the practice which shows the deviation from this conception. What will happen here? Høg believed there will be only one Fourier term in the flexure, with no greater coefficient than $0''.02$ – which has to be determined in the conventional way. Klock pointed out that there are new methods available for determining the instrumental constants, giving as example Metič's work at Belgrade with a new enclosed tube mark system. Van Herk recalled in this connection the laboratory work done in 1953 by him and De Munck with a tube of 23 m length kept at an inside pressure of 1 mm.

An argument for the need to improve upon instruments was given by VAN HERK who tried to explain Oort's empirical result of 1943, in which he found the technique of observing members of the solar system to improve the equator point in the average to be of the same value as the average from a sufficient number of catalogues where no such observations of solar system members were performed. The systematic errors made in the observations hamper the outcome for the equator point and the equinox. In particular the terms $\Delta\delta_\alpha$ and $\Delta\alpha_\alpha$ will turn up in the computed 'corrections'. Tucker suggested it is not so much a better type of instrument we need, but a better type of observable object, one which will not show such large seasonal correlations which are inseparable from Sun observations. Stoy was of the opinion that systematically we have practically reached the limits of our instruments, and saw the cure as having more instruments to average the errors out. Klock wanted to know whether there is a limitation of declination errors $(\Delta\delta_\delta)$ with respect to refraction. Nobody offered an answer. Reiz was of the opinion that, with the high accuracy with which division errors can be determined today $(\pm 0''.02 - 0''.03)$ it should now be possible to reduce the systematic terms $\Delta\delta_\delta$ considerably, perhaps even to eliminate it completely. Høg disagreed here: most of the $\Delta\delta_\delta$-term comes from flexure in the tube, the objective, and the micrometer and from the uncertainty in the refraction constant, which in turn is again insufficiently known through the error in the flexure. Gliese reminded us of his *Astron. J.* 1965 paper in which he showed the existence of fluctuations of instrumental systems which cannot be explained.

Of more future plans, LEVY gives an exposé of the efforts made in France where, after a comparison of 6 possible sites for a new astrometric station, the Plateau de Calern, north of Grasse (AM) has been singled out. The site (6 km × 2 km, altitude 1200 m) will be the future location of CERGA (Centre d'Études et de Recherches

Géodynamique et Astronomiques). Since 1971, a time service and a Danjon astrolabe have been installed. There will ultimately be a meridian circle, a wide field astrograph, a long-focus astrograph, equipment for observations of satellites, equipment for distance measurements, a long base line interferometer, gravimeters, and clinometers to study the inclination of the lithosphere. Instruments and workers from foreign countries will be welcome.

REQUIEME described the Bordeaux photo-electric micrometer. A rotating edge-sensor performs the centering on the star image, as in some self-guiding telescopes. The following standard errors are computed for one observation: $\varepsilon_\alpha = \pm 0\overset{s}{.}0069$ (reduced to the equator) and $\varepsilon_\delta = \pm 0\overset{"}{.}20$, with $m < 9.2$. For PZT stars in the declination zone $+40$ to $+60°$, these results are still better ($\pm 0\overset{s}{.}0059$ and $\pm 0\overset{"}{.}168$). A good mounting of the machinery is absolutely essential. The difference in accuracy between α and δ is related to the influence of the divided circle needed in the determination of the δ. Høg was worried over two things: (1) the atmospheric dispersion is a serious problem and with no filter being used to limit the observations to a sufficiently narrow band, the results will show systematic differences on the colour of the star (at $z = 60°$, a B and M type star will show a systematic difference of $0\overset{"}{.}2$, an error very dangerous for astrophysical applications. A visual colour filter should be used, which will cut off at least 1 mag.; (2) the limiting magnitude at dark sky is quoted as 9.5 – with colour filter 8.5, which is 2 mag. less than the Hamburg-Perth instrument has performed. Strömgren had mentioned the possibility of reaching 13.5 mag. Where lies the reason for this discrepancy? The answer is to be sought mainly in the use of too large a diaphragm ($1\overset{'}{.}5$) so that too much sky background and faint stars increase the noise level. Strömgren remarked that we have learned from photometry that dark current is no longer a problem. With a good photo-multiplier and pulse counting technique, even at ambient temperature, the dark current should be 100 times smaller than quoted here. The mechanical qualities of the set-up were so good that it would be worthwhile to try a much smaller diaphragm. Requième pointed out that the size of the latter is imposed by the difficulties of setting the telescope correctly, but Strömgren expressed the opinion that a telescope can certainly be set to a few seconds of arc, and a diaphragm of $15''$, as in photometry, should be sufficient. At the request of Benevides-Soares, the time constant for the servoloop and the time interval between contacts at $\delta = 0°$ ($0\overset{s}{.}7$) were given. Klock asked if stars close to the pole could be tracked, and this was confirmed.

LACROUTE proposed the combination of results from two different types of instrument, the meridian circle and the Schmidt camera, in order to connect the system of bright stars, usually observed with meridian circles, with the fainter stars used in connection with the galaxies. The positions from the Schmidt camera come from the over-lapping field techniques. He set out from adopted values for the accuracy with which a meridian observation is carried out (taking Høg's values) and for a Schmidt camera position, using the values given by Andersen. He suggested that each star should be observed with the meridian circle three times to eliminate discordant results. The Schmidt camera positions, though apparently suffering from important systematic errors in some locations on the plates (larger than the purely accidental errors) are in the average of fair accuracy. The Schmidt camera will carry a picture of a reseau to check the relationship between the curved plate and the plane field. The fundamental system (FK4 or a revised improved version) has to be tied by the meridian circle to faint stars (down to $m = 11$), and these faint stars have to be tied to the photographic catalogue of faint stars. What can be achieved in this respect depends on the size of the field in the sky over which the systematic error can be considered to be constant, depending on the average star density in the field, and on the accuracy of the positions. His methods explained in *Ann. Strasbourg* **VI**, 1964, are used. With estimated values for the parameters mentioned, two schemes are considered – a minimal and an optimistic – and the numbers are now given for the stars for which the positions have to be determined in order to reach a required accuracy. In the discussions Høg reported that Dieckvoss has come to the conclusion that the big fields from Schmidt cameras cannot be used for astrometric purposes. Perhaps a field flattener could help. Fricke and Van Herk stated that the Schmidt will give good results as long as the portions of the plate used are small. Strand pointed out that this is different from the problem Lacroute has discussed, and he feels strong doubts about the feasibility of the Schmidt for astrometry over an extended field. Reiz referred to Dixon and Andersen's work and believed that, with the precaution taken to use a circular ring as diaphragm as a mask plate pressed against the photographic plate, the proposal of Lacroute would be possible. Lacroute added that the errors arising from the use of different parts of the mirror could be improved upon by using new non-dilating glass for the mirror.

LOPEZ reported on the Felix Aguilar observatory situated 10 km west of the city San Juan. It is run through an agreement between the Universities of Cordoba and Cuyo. A Repsold meridian circle, with a motor-driven micrometer, and a Danjon astrolabe have been in operation since 1968. Up to August 1972 some 40 000 observations on 7190 stars have been made with the meridian circle (the reductions following closely) and 35 000 observations on over 300 stars with the astrolabe. Differences in positions with respect to the FK4 have already been given in Brighton.

BENEVIDES-SOARES gave an exposé of the Brasilian meridian circle at Sao Paolo. The instrument is of the Askania type, built by Zeiss (Oberkochen). Great care has been exercised to ensure high mechanical and thermal stability. The pillars are cast from a very homogeneous mixture, anchored solidly to the granite base. The building is protected against solar radiation and has a low heat capacity. Ventilation during the day is applied. The instrumental errors have already been checked regularly; the inclination shows some instability. The impersonal micrometer is motor-driven, the δ-micrometer readings are printed. The 40 cm glass circle will be read by an automatic digital read-out system. A fundamental program of FK4 and FK4 Sup stars is planned.

DEBARBAT recalled various favourable opinions on the merit of observing positions of major and minor planets, and mentioned their importance for theoretical investigations. The Danjon astrolabe is very well suited to participate in this type of work alongside the meridian circles, and in fact the observations over the years 1965–67, made by the two types of instruments do not show any substantial systematic difference, which is very gratifying. It is for this reason that an opinion is asked about a proposal to be made to Commission 8 to recommend the use of Danjon astrolabes for future observations of planets. In the discussion, the unfamiliarity of the practical side of the problem gave rise to doubts whether the connection between the astrolabe observation of the planet and the fundamental stars will be as good as is believed to be the case with the meridian circle observations. In the latter case, from 8 to 30 fundamental stars are observed together with a planet. Now, the astrolabe observers use, in a $1\frac{1}{2}$ h tour, 28 fundamental stars, which are then tied to a planet.

2. Reference Systems

The problem of the fundamental system was broached by FRICKE. He referred to his latest papers: 'Prospects of an Extension of the Fundamental Reference System to Faint Objects and Radio Sources', in press in *Mem. Soc. Astron. Italiana*, and 'Fundamental Systems of Positions and Proper Motions', in *Ann. Rev. Astron. Astrophys.* **10**. The principles underlying the construction of a fundamental system were shortly described. If all the difficulties connected with the motion of the Earth and motions of the stars were overcome, we should have a non-rotating reference system, and the question of the differential galactic rotation would pose no difficulties. All this is not the case in practice. Radio observations can determine today the position of the equator with a systematic accuracy reached in optical astronomy in 1820. There is, however, good reason to believe that the accuracy in absolute declinations will soon be the same as with modern meridian circle work. It is worthwhile to look for the optical counterparts of the radio sources. The α's of these optical counterparts have to supply the zero point in α for the radio system. Proper motions could be improved by connecting them with galaxies. The p.m. obtained via galaxies are free from inaccurate definitions of the equator and equinox, and from uncertainties in the precession. Though the Pulkovo and Lick observatories have already worked in this direction, the goal has not yet been reached. The proper motions with respect to the galaxies agree well in declination between the two observatories, but in α there exists an unexplained difference of $1''2$ per century. In the Lick motions a magnitude effect appears to be clearly indicated, and some investigators want this difference to be explained as a real rotation of the brighter stars with respect to the fainter ones around the axis of rotation of the Earth.

Good proper motions alone would not be of much help for a fundamental system, as no ephemerides can be provided in that case. What about a possible improvement and extension of the FK4? There is no doubt the FK4 should be extended to objects as faint as 9^m. However, these objects can only be picked out after careful consideration of their suitability: that is, a good proper motion must be found as otherwise the star is of no use in a fundamental system. This depends on the observational history of the star. Perhaps the star list prepared in Leiden on the basis of observational history could be of use here. The stars may not be chosen in a narrow range of spectral type – they have to represent all types equally well. In the discussions on this matter, Fricke pointed out he is not in the position to say at this moment which stars, or how many stars will be included in the selection. This has to be considered carefully. The steps to be taken can be arranged under the headings:

(1) Documentation of the observational data;

(2) Problem of extension to fainter limit;

(3) Systematic improvement of the FK4;

(4) Individual correction to data;

(5) The question as to the fundamental constants to be used in defining the new system.

It is emphasized that communication with the observers is very necessary, as past experience has shown that much valuable information cannot be obtained in any other way. The systematic relation between the catalogues and the FK4 has to be investigated. Sometimes incorrect relations are given, so this has to be checked. Therefore it is urged that the results of observations on FK4 stars are given and published by the observers. With respect to the extension to the 9th mag., it was once suggested to use the General Catalogue by Boss, and revise this as soon as enough observations on faint stars would be available. But the Boss system shows a large magnitude error, and the idea had to be abandoned. Fricke wants to use the AGK3R and the SRS as a basis instead, to pick out the fainter stars to be included in the FK4. In building the new system, observations from prior to 1900 will probably no longer be used. For the documentation Gliese will give a report.

As to matters of a general nature, the equator is more easily determined than the position of the equinox. The equator for the FK4 was derived from observations of the Sun and planets from 1857 to 1949. The corrections to the FK3 equator were small quantities within the range of the formal accuracy. Leaving out catalogues from before 1900, the same result was obtained. It is doubted whether 20 more years of observing will improve the situation materially. With the equinox the situation is different. The corrections found to Newcomb's N_1 gave values between $-0\overset{s}{.}028$ and $-0\overset{s}{.}073$, but no secular change could be detected. The fundamental p.m. in α are affected by an erroneous motion of Newcomb's equinox; the correction, according to Fricke, to be applied to the FK4 p.m. is $1\overset{''}{.}23 \pm 0\overset{''}{.}15$ per century. The inconsistency in the FK4 (present in all catalogues after 1900) is that no attempt has been made to eliminate this effect in the p.m. – only regional corrections were applied. The same is the case with the obliquity of the ecliptic. A complete rediscussion will be made to investigate the ΔE and $\Delta \varepsilon$ plus new data on the precessional constant. The compiler is not free to use the constants he deems best, but has to stick to the ones which have been internationally agreed upon. It is pointed out that the difference found in the precessional constant is caused by the difference between the declination system of the first fundamental catalogue and the declination system of the FK4. Newcomb took the Auwers-Bradley catalogue of p.m., improved this a little, and the difference between this system and the modern catalogue system is a rotational component of the declination p.m. system around an axis directed to 6^h. Personally, Fricke believed it is time to change the precessional constant, but it is necessary to consider all consequences. A change is likely to be made with the adoption of new masses of the planets which play an important role in constructing ephemerides. Perhaps a decision can be taken in 1976.

Tucker asked whether the criteria of observational history cannot be relaxed so that a more restricted range of magnitude and colour may be represented in the faint stars added to FK4, but Fricke was against this restriction. Schombert reported that within 6 months the USNO will have available five or six thousand proper motions of AGK3R stars with the best observational history. Klock announced that the 6-in. transit circle at Washington has added Flora and Metis to its minor planet list. The major planets are now observed over a much wider range of arc. Van Herk remarked that with the change in p.m. system from FC to FK4, a change in the precessional constant was to be expected. But it remains necessary to explain why this big change did take place, there is something to worry about. Fricke stated that the NFK contains results of many more observatories than FC did, and the position of the equator was, for the first time, determined from solar observations in the northern and the southern sky.

GLIESE reported about the documentation of catalogues made available in the last 20 years, and which were not yet incorporated in the FK4. A preliminary inspection dealt with the instrument, the program, the epoch, region of δ, number of stars, the magnitude range, the observational errors, and whether the catalogue was observed fundamentally or not. From 38 observatories a total of 143 catalogues is available; of these, 102 include bright stars $(m < 6.5)$, and 89 include faint stars $(m > 6.5)$; 107 give α-observations, 81, δ-observations (the time services are responsible for this unbalance). In an optimistic mood, Gliese guesses the number of absolute catalogues to be 25. In all some 140000 positions of stars are given. The PZT, the Latitude and the Astrolabe lists will be investigated. Some catalogues are published in combinations, and it will be necessary to investigate how these will be dealt with. References to observations which will be finished in the next years will be very much welcomed. Often Gliese finds in literature comparisons given of observations with a fundamental system, but the conditions of observing are often not given. To judge the catalogue, these details should be published. Stoy wanted to know whether planets should be observed every night. Fricke, referring to the work of Jackson at the USNO, said it does not help to observe the planets every night, as long as the observed arcs remain short. Strömgren was of the opinion that all minor planets should be observed.

Stoy advocated the use of the results from photographic catalogues as well in constructing the fundamental system. Fricke was very doubtful in this respect as he fears the magnitude errors in the photographic catalogues above all. Van Herk ventured to remark that meridian circle catalogues show magnitude errors as well. He had, however, an example of a large discrepancy in photographically determined p.m. similar to that mentioned by Fricke: in the work done on more than a thousand common stars at Hamburg (Osvalds) and at Leiden (Pels), the p.m. in δ agree reasonably well, but the p.m. in α show systematic differences of the order of $1''$ per century, totally unexplained.

The state of the internationally executed reference star programmes was described by SCHOMBERT for the epoch July 1, 1972. His report culminated in a table in which, for each participating observatory is given, amongst other items, the declination zone to which each instrument was committed, the number of stars to be observed, and the number of observations to be performed. This table will appear in the IAU report of Commission 8 and is, therefore, suppressed here. The number of observations carried out is very near to 100%. Of the reductions necessary (it may be recalled that the USNO carries out the apparent place reductions for all participants who wish it, and a great deal of the refraction computations as well), an average percentage of something near 78 has been finished.

The question how the final catalogue of the SRS stars will be compiled, drew some debate. Anguita asked whether the final reductions of the SRS stars should be done in the original FK4 system or, knowing that there are strong systematic and accidental errors in the FK4 in the southern hemisphere, should the reductions be done in an improved FK4 system? Gliese remarks here that various published observations show serious errors in the southern α-system in the FK4. More observations are forthcoming. The question arises whether an intermediate α-system should be compiled or, whether we should wait for the results of the observations now under way? Anguita mentioned more recent results than Gliese did. Fricke was of the opinion that the SRS stars should be reduced in the FK4 system. Høg pointed out that this is not an unambiguous statement. Every observatory has its own tradition for distributing the FK4 stars during the night and along the meridian. This is one reason why each observatory ends up with its own system. Every observatory should use its own improved FK4 positions if they wish, provided the observed positions of the fundamental stars are published. This is an essential requirement which has not always been fulfilled. Anguita framed his question in a slightly different way: Will the final SRS catalogue be given in the FK4 system or in some improved system? He further raised the question who was in charge of this program at the beginning. There had been a division of the work at the $-30°$ boundary, one zone under the responsibility of the USNO, the other under the Pulkovo observatory. Strand recalled the past history of the program, the selection of the stars, the chairmanship of the subcommittee of Commission 8, and what part the USNO has done in the reductions. Schombert reported that an agreement had been made between Pulkovo and Washington: all observations will be reported to the USNO and to Pulkovo, the first undertaking to compile the final catalogue. The proposed resolution was amended at the request of Tucker, so that other institutions would be entitled to receive the data.

YASUDA dealt with the special programmes. He pointed out the importance of programmes like the PZT star programme, as it is only possible to connect the different zones of PZT stars with a meridian circle.

It is hoped that, in doing so, systematic errors in the different zones which are, at present still there, will be attenuated. This would result in a better control over the polar position, as the latter is connected with errors of the type $\Delta\alpha_x$ and $\Delta\delta_x$. The secular variation of the polar motion is affected also as the p.m. of the PZT stars are uncertain. The observational part of the northern PZT star programme is shared by 12 meridian circles and one vertical circle, with 4–6 observations per star for each instrument. The results from the observations performed between 1950 to 1970 will be compiled into a preliminary catalogue, and each time a new complete set of observations becomes available, this set will be incorporated in the existing catalogue.

The special programmes on O–B stars are observed at Tokyo. It was stressed that these stars, as well as Cepheids should be observed in the south. Tucker expressed disagreement with Yasuda's phrasing of the recommendation to limit the observations to narrow zones. The main object, to link the different zones, will be endangered. Requième remarked that, for good differential observations, narrow zones would be best, but to link the zones one should go to at least 40° width. Lacroute agreed with this suggestion, as it all depended on what purpose one has in mind. Gliese asked how the stars would be chosen to make the connection between the different zones. Høg pointed out the importance of having all division errors determined in work on narrow zones. Dejaiffe recalled the possibility of linking the PZT stars to the FK4 through the intermediary of the ILS catalogue, which has been tied to the FK4 by the Melchior-Dejaiffe catalogue. It would be profitable to include more ILS stars in the PZT programmes to improve the interconnection.

3. Various Investigations

ISOBE had analyzed the results of the observations of FK4 stars obtained during the AGK3 campaign of 10 meridian circles, and of those stars observed with the Tokyo m.c. during the SRS program. He separated the influence of the systematic error from that of the random ones by analyzing the $(O-C)$'s according to the number of observations on each star. The equation of condition is the well-known relation between the variances for the total error, the systematic error and the nth fraction of that for the random errors, with n the number of observations on the star. He concludes that, in general, with a number of 10 observations per star, the influence from the random errors is of the same order of magnitude as from the systematic errors. Notable favourable exceptions are the Nikolajev and Washington 7-in. instruments.

SCHOMBERT presented the analysis of the differences of the AGK3R stars as found in the AGK3 with those given in the meridian work. There is a discrepancy between the two catalogues in α. These differences exist for a number of consecutive hours, and then vanish. For different zones of δ, these peaks shift to slightly different hours. These differences show a small scatter, and have, for nearly all 5° zones the same value of approximately $+0''.09$ (sense AGK3R–AGK3). These differences were noticed when the proper motions of the AGK3R stars were derived (the mean errors of these p.m. are of the order of $\pm0''.35$ per century) and these p.m. showed fairly systematic differences from the AGK3 p.m., being large near the equator and in the late hours. It was remarked by various people that the differences just described could, perhaps, be due to magnitude errors, and Schombert will look into this matter further. (At the time of the colloquium this investigation was not yet complete. Further investigation has revealed that the differences mentioned are no longer significant: application of the influence of the proper motions due to the differences in epoch between the meridian and the photographic observations made the differences in position disappear. The findings are now in good agreement with the results found by Dr Dieckvoss earlier).

MARCUS discussed the 3000 differences in positions of the Bucharest KSZ stars with those given in the AGK3R. In the equatorial zone of 2° width, the greatest value for $\Delta\alpha$ is $+0^s.010$ for $10\,h < \alpha < 12\,h$. This is similar to the findings given in Schombert's work. – In the discussion: Høg was of the opinion that the similarity with Schombert's work is spurious, and Schombert was afraid the AGK3R could be wrong. – The majority of the $\Delta\alpha$ differences are positive, that of the $\Delta\delta$ negative. A magnitude relation seems hardly to exist. Graphs, representing a three-dimensional representation of the differences show the $\Delta\alpha_\alpha$, $\Delta\alpha_\delta$, $\Delta\delta_\alpha$, $\Delta\delta_\delta$ relations. Küstner series of FK4 stars served to correct the Bucharest observations to the FK4 system; for the declinations, no such correction was found necessary. The differences have been represented by trigonometrical expressions, with coefficients determined from the maxima and minima of the surfaces, using the second derivatives to determine the signs. In the discussion, Marcus points out that temperature differences have been measured at Bucharest, where the meridian circle is poorly situated with respect to the surrounding buildings. Lacroute advocated the method of 'Synthesis' used at Strasbourg to get the best possible connection with a fundamental system. This method reduces the random errors at the same time.

TELEKI sees great need for a more detailed study of the astronomical refraction. The study group, established by the IAU Commission 8 has made a number of studies in this field. Variations in the air density around an observational site depend on the time of the day, the amount of water vapour, and the refractive index cannot be calculated, at present, with sufficient accuracy, neither with respect to accidental, nor to systematic errors. This is a consequence of insufficient knowledge of the meteorological elements, and of the lack of dynamical parameters in the formulae. The vicious circle formed by the dependence of the refraction determination upon the instrumental flexure should be disrupted. The atmospheric dispersion as well as the inclination of the lower layers of the atmosphere should be taken into account. Studies on an extension of the theory to an asymmetrical distribution of the air-density are under way. The almost universally used Pulkovo tables should be revised for every observatory to meet its local circumstances. Plans were outlined to connect the refraction tables to aerological data, to study the non-homogeneous refraction, depending on the size of the scattering particles, the influence of the absorption lines. A full report will be given at Perth.

Høg's request to specify what an individual observer has to do, could not be specifically answered, but contacts with the nearest meteorological station would help. Tucker thought it preferable to determine corrections to the standard tables as is done at Greenwich, rather than introduce a greater number of standard tables. Teleki emphasized again that refraction is a local effect. The determination of the correction for refraction directly from the observational data is not correct. Strand recalled the work he had done on determining refraction near the horizon with the aid of lunar photographs (Dearborn), and he found great changes from day to day with systematic differences of up to 5%. Fricke suggested that, as radio astronomers have difficulties in determining the refraction, a radio astronomer should be contacted for the study group. Both Fricke and Van Herk stressed that the colour differences amongst the stars had a non-negligible effect on the refraction.

VAN HERK described some results of catalogue comparisons. A magnitude term is still demonstrable, and it would be better if meridian circle observers published the value of this parameter, the true apparent magnitude while observing. He defended on physical grounds the process of determining the $\Delta\delta$ according to α directly: stars at 5 h right ascension are observed in entirely different conditions from those at 17 h. This can be shown to be the case in practice. A two-dimensional plot of the differences show areas of equidifference provided the average star density is not too high. In regions where this density is much higher, this gentle picture is disrupted. It seems obvious that, with many more stars, many more nights are involved, with the consequence that entirely different systematic errors traverse each other. The influence of the differences found between the first Grw 1950 and all other catalogues was mentioned; here it is seen that within a small zone, of only 6 to 8 deg width, large differences can occur. Tucker stated that at Herstmonceux, no differences are found in results for stars observed with and without grating. Reiz asked whether the Brorfelde meridian circle, assuming that this is working without magnitude errors, could be of help in checking this type of error in other catalogues, as there will soon be available a catalogue with some 3000 stars in it. Van Herk hoped that the results on the observed FK4 stars will be published as well, as otherwise the comparison would be much more difficult. Gliese asked for an explanation for the fairly large magnitude equation observed between the photo-electric positions in the Russian Time Service Catalogues and the fundamental catalogues N30, FK4. Høg replied that the reductions for his multislit micrometer are entirely symmetric, but in the Russian method one has to rely on a time constant of a d.c. amplifier, and thus a magnitude term could become possible.

YASUDA described the comparison of the 8 PZT catalogues now in use with the AGK3R; the positional differences are analyzed according to $\Delta\alpha_\alpha$ and $\Delta\delta_\alpha$, and the periodic coefficients computed. A large dispersion exists amongst these coefficients. The observations made differentially to the FK4 system have revealed already differences with the FK4 of the type $\Delta\alpha_\alpha$ and $\Delta\delta_\alpha$.

HØG recalled that we are now fairly sure to be able to observe very faint stars with meridian circles in the near future and saw the problem arising of the choice of a reference catalogue of faint stars, to which Fricke emphatically agreed.

MARCUS mentioned that the Prague Observatory would like to see their list of PZT stars embodied in the general programme of PZT stars observed by meridian circles. FRICKE pointed out again what is needed: assume an O star to be at 1 kpc distance. The cosmic scatter in the radial velocity is of the order of 10 km s^{-1}. To reach this limit in the proper motions, these should be known far better than what is achieved today. The idea that the AGK3 would help in this respect was unfounded optimism – we need p.m. with a standard error at least 3 times smaller. Therefore, many more meridian circle observations on O and B stars will be necessary to achieve this goal.

VAN HERK read a letter by Dr J. DOMMANGET, President of Commission 26 (Double Stars), in which the participants are reminded of the urgent need for meridian observations of visual binaries, resolved or unresolved (IAU Colloquium No. 18). If any meridian circle observer is interested in the subject, members of Commission 26 offer their help in establishing suitable programmes.

MARCUS mentioned the observation of double stars at Bucharest; Van Herk recalled the programme started at the USNO in the sixties. Strand mentioned a number of 700 double stars which have visible motion. Reiz was of the opinion that it was not worthwhile observing too narrow double stars with an instrument with a focal length of 250 cm.

4. Resolutions

The following resolutions were brought forward to be recommended to Commission 8 of the IAU.

(1) The recommendation made by Commission 8 at the XIV General Assembly concerning the development of new instrumentation and techniques is re-affirmed. In particular, the provision of a modern transit circle for the Tokyo Astronomical Observatory is strongly recommended. It is further recommended that a Study Group be established for horizontal meridian circles.

(2) It is recommended that an improvement of the FK4 and its extension to a fainter magnitude limit, resulting in a new fundamental catalogue, the FK5, be carried out at the Astronomisches Rechen-Institut, Heidelberg; that observatories throughout the world contribute to this project by providing basic observations, on punched cards if possible; and that all information pertinent to the formation of the FK5 be transmitted to the Astronomisches Rechen-Institut with the observations.

(3) It is recommended that meridian catalogues should include, in addition to the tabular magnitude on a recognised system, an indication of the screens used for each star, so that the approximate magnitude at which the star appeared to the observer can be deduced.

(4) It is recommended that programmes of observations of fundamental stars, with meridian circles and astrolabes, should include members of the solar system.

(5) It is recommended that the SRS catalogue should be compiled by the U.S. Naval Observatory, Washington; and that this observatory should send the basic data to the Pulkovo Observatory and, on request, to other institutions.

(6) The recommendation made by Commission 8 at the XIV General Assembly concerning the inclusion of FKSZ stars in fundamental observational programmes is re-affirmed, with particular reference to observatories in the southern hemisphere.

(7) It is recommended that the work of the Study Group on Astronomical Refraction that has been formed by Commission 8, should continue to be supported.

DISCUSSION

Fricke: I should like to make an addition to Dr van Herk's report. The resolutions adopted at Copenhagen have been submitted to be considered and endorsed by Commission 8 at Sydney. As I have found out in correspondence with members of the Commission, these resolutions will require some changes and additions in order to obtain the approval of the Commission. The most important change concerns the compilation of the SRS catalogue. The compilation is to be carried out for the whole southern sky at the Pulkovo Observatory and at the U.S. Naval Observatory. Consultations on the best possible system for the SRS catalogue shall take place between the Astronomisches Rechen-Institut at Heidelberg, the Pulkovo Observatory and the U.S. Naval Observatory.

PLANS FOR THE IMPROVEMENT AND EXTENSION
OF THE FK4

(Invited Paper)

WALTER FRICKE

Astronomisches Rechen-Institut, Heidelberg, F.R. Germany

Abstract. Recent absolute and differential observations of fundamental stars have indicated the need for a revision of the fundamental reference system represented by the positions and proper motions of the FK4. The revision of the FK4 will be based on modern fundamental and differential observations of high accuracy which allow (a) the extension of the FK4 to fainter stars, (b) the improvement of the system, (c) the derivation of individual corrections to positions and proper motions, and (d) the determination of the equinox and its motion. The completion of the programme, resulting in the FK5 and a Supplement giving data in the system of FK5 for bright stars and faint objects (including optical counterparts of radio sources), may be expected about 1980.

1. Principles of Establishing a Fundamental Frame of Reference

The Fourth Fundamental Catalogue (FK4) represents the conventional fundamental frame of reference adopted by the IAU in 1961 (*IAU Trans.* **10**, 79 and **11B**, 1967). This catalogue gives the positions and proper motions of 1535 stars that are fairly uniformly distributed over the sky and are brighter than visual magnitude 7.5. For practical reasons the system is that of the Earth's equator as the fundamental plane from which declinations are reckoned, and the vernal equinox as the origin for reckoning right ascensions. The system is established by absolute observations of positions of stars and members of the planetary system at various epochs.

Any approach to improve the fundamental frame of reference has to begin with a critical survey of the possibilities for a revision of the principles underlying the construction of the reference frame. The choice of the equatorial system is loaded with a number of difficulties due to the orbital motion of the Earth and to the motion of the equator and equinox. It is therefore appropriate to ask whether any other choice could be made. Well-known are the advantages of an extragalactic reference frame represented by positions of galaxies in a net-work over the sky. The net of positions can be established, in principle, without relation to the equatorial system, and a zero-point can be arbitrarily chosen. In practice, however, we are still far away from a realization of this possibility, at least, as far as the independence of the net of extragalactic positions from the equatorial system is concerned. In an unforeseen way the new radio-interferometer techniques which provide declinations and right ascensions of radio sources without reference to previously determined positions, support the continued use of the equatorial system as the proper choice for the near future.

There are other important arguments in favour of no change in the principles of establishing a fundamental reference frame at present, but in favour of a drastic improvement of the conventional system on the basis of available observational ma-

Gliese, Murray, and Tucker, 'New Problems in Astrometry', 23–30. All Rights Reserved.

terial by means of classical methods and by all possible attempts at relating the system of bright stars to the radio astrometric system. First, I may mention that the FK4 can hardly serve its purposes satisfactorily for more than about 20 years from the date of its completion. With advancing epoch of observation, the accuracy of star places computed from the FK4 decreases, mainly due to the errors of the proper motions. Second, the increase in the accuracy of modern observations and an increasing diversity of applications demand for more reliable ephemerides. Third, during the past two decades a large number of absolute and differential observations of FK4 stars and fainter stars have been made with improved techniques with the classical instruments, the meridian circle, vertical circle, and Danjon's astrolabe, such that it is justified to start with the improvement of the FK4 and of extending it, hopefully, to the ninth visual magnitude in a new fundamental catalogue, the FK5.

The work before us consists of several steps, most of which are closely related to each other and must be done one after the other, and some are independent from each other and can be done simultaneously. Even with a perfect organization of the work, the completion of the FK5 can hardly be expected before 1980. The various steps may be described by the following main headings:

(1) Documentation of basic observational data and study of the techniques that have been applied;

(2) Extension of the FK4 system to a fainter magnitude limit;

(3) Derivation of systematic corrections to the FK4 from absolute observations;

(4) Derivation of individual corrections to the star places and proper motions in the FK4 from absolute and differential observations;

(5) Improvement of precessional values and elimination of errors due to incorrect motion of the equinox (a change in the conventional values of precession requires international agreement);

(6) Establishment of a Supplement to the FK5 giving accurate data of bright stars and faint objects including optical counterparts of radio sources.

I will proceed now in describing various aspects of these tasks in more detail except item (5) concerning precession which will be one of the general themes of the Joint Discussion No. 1 at the XV IAU General Assembly.

2. Documentation of Observational Data

The documentation of observations relevant to the improvement and extension of the FK4 includes the punching of the data of a great number of observational catalogues; general catalogues which are the result of a compilation of different sources will not be used. We are dealing with two groups of catalogues, the first consists of those since about 1900, in which the observers have included stars fainter than visual magnitude 7.5, and where measures have been taken to avoid errors depending on the magnitude. These observations will contribute to the extension of the FK4 to fainter stars. The second group consists of catalogues which have become available since the completion of the FK4. They may contribute to the improvement of the systematic and individual accuracy, and, of course, also to the extension of the FK4. The total number of catalogues which will have to be prepared in a form suitable for data processing will be between 160 and 200. Many of these have already been punched in a joint project of the U.S. Naval Observatory and the Yale Observatory and others at Heidelberg. In addition, some observatories have provided their catalogues on punched cards, but these are favourable exceptions to a rule which in future, should be followed. Together with the documentation of observations goes the study of the methods, which have been applied by the observers, in order to decide for which purposes the material can be used. In fact, it is often not easy to find out to what extent observations are absolute, quasi-absolute, or differential to a system which can not be expected to be the system of the FK4.

We expect that work at the documentation will take us another few years, and that even after its completion we should be prepared to include in the program data that may become available within the next three years from now on. This implied a continuous contact between the observers and the compilers of the FK5 at Heidelberg.

3. Extension of the Fundamental System to Fainter Stars

At the time of the compilation of the FK4 the available observations were not sufficient for an extension of the system beyond the visual magnitude 7.5 without considerable loss in systematic and individual accuracy. Almost all observations made before 1900 are seriously affected by errors depending on the magnitude, and even after 1900 it has taken some time to convince observers that provision for the elimination of 'magnitude equations' in declinations and right ascensions is necessary. During the past two decades the situation has considerably improved. Furthermore, the international reference star programs AGK3R (reference stars for the AGK3) and SRS (southern reference stars) have provided observations down to about ninth magnitude which are either strictly in the system of the FK4 or in instrumental systems whose relations to the FK4 are well determined. There are about 40000 stars distributed over the whole sky in the reference star programs. The observations in the AGK3R were shared by ten observatories and completed in 1963. The resulting accuracy of $\pm 0\overset{s}{.}005$ sec δ and $\pm 0\overset{''}{.}116$ in right ascension and declination is excellent. One may therefore consider the AGK3R as the most comprehensive sample of stars in the northern sky suitable for the selection of stars down to ninth magnitude for the extension of the fundamental system. At Heidelberg, we have adopted the AGK3R as the 'master catalogue' from which stars in the magnitude interval from 7.5 to about 9 will be selected for inclusion in the FK5. In the southern sky, the SRS catalogue shall serve the same purpose. The selection will depend on the observational history of the stars, and the final choice will have to be made on the basis of the accuracy of proper motions that can be derived from all suitable observations made from about 1900 onwards. Concerning the spectral composition of a fundamental catalogue, I am of the opinion that the distribution should not be restricted to a narrow range in spectral classes. How useful it would be, if at least a few stars brighter than ninth magnitude with radio emission could be included in the FK5! The advantages of such stars for linking the radio astrometric system to the fundamental system have only recently become apparent.

Besides the primary criterion of accuracy of positions and proper motions, secondary criteria may be the distribution of the stars in the sky and in distance from the Sun. For practical reasons, the representatives of a fundamental reference frame should be distributed fairly uniformly in the sky. On the other hand it is attractive to include in the efforts of deriving proper motions of highest systematic precision as many stars as possible at large distances from the Sun. This aim implies that MK spectral types and accurate photometric data are available.

Besides the AGK3R and SRS catalogue as sources for the extension of the fundamental system, the FK4 Sup catalogue will provide additional material. FK4 Sup stars have been included in many observational programs, and many of them will fulfil conditions of sufficient systematic and individual accuracy for inclusion in the FK5. Although it is not possible to make a well-founded estimate of the number of stars in the FK5 before the investigation of the observational history of the stars in the AGK3R, the SRS catalogue and the FK4 Sup will be completed, I expect that the number of 1535 stars in FK4 will have to be increased to at least 3000 and at most to 5000 in the FK5.

4. Improvement of the System and of the Accuracy of Individual Data in the FK4

Concerning the prospects of an improvement of the system of the FK4 and the derivation of individual corrections I refer to the report to be given by Dr Gliese except for a few questions of general nature. These questions concern (1) large scale inhomogeneities in the system over the sky, which have become apparent by systematic deviations of modern positions from the FK4 in various regions of the sky, and (2) errors in the determination of the equinox and the equator point. Among the inhomogeneities appear to be most prominent the errors in right ascensions in the zone $-50°$ to $-80°$. They were found by the joint effort of Chilean and Russian astronomers working at different instruments at the Santiago Observatory, and they were confirmed by other observers as well. If one takes into account that the FK4 system south of $-30°$ rests mainly on observations of a single observatory, the Cape observatory, such errors are not surprising. It is more surprising that no serious errors in the southern declination system have become known. The absolute observations made during the SRS program are, however, not yet completed, and the final derivation of corrections to the FK4 system in the south must wait for the results which may become available within the next few years.

The equator point and the equinox are being determined by observations of the Sun and other bodies of the planetary system. In fact, almost the same series of observations contribute to the determination of both, and they also yield corrections to the obliquity of the ecliptic. The results depend mainly on the absolute measurement of declinations. Such measurements determine more easily the position of a great circle in the sky, the equator, than the position of the Sun in the equator and in the solstices. From the results which we have derived for the position of the equator in the FK4 – these results have not shown any secular change – I do not expect that more recent observations will give us an appreciable correction. The situation is different with the determination of the equinox and the obliquity. The position of the equinox in the FK4 is defined by $N_1 + \Delta N_1$, where N_1 denotes the equinox in the right ascension system of equatorial stars in the *Catalogue of 1098 Standard Clock and Zodiacal Stars* by Newcomb (1898). In the same year this equinox was applied by Newcomb in the Catalogue of Fundamental Stars. Later, appreciable corrections ΔN_1 were found from the observations of the Sun and other bodies of the planetary system. All observations carried out from about 1900 to 1950 have provided values of ΔN_1 in the interval

$$-0\overset{s}{.}028 \geqslant (\Delta N_1)_{\text{Cat}} \geqslant -0\overset{s}{.}073$$

with no indication of a significant secular change. In the FK4 the correction ΔN_1 $= -0\overset{s}{.}050$ was adopted but it has not been taken into account in the proper motions. In fact, all compilers of fundamental catalogues after Newcomb have avoided to eliminate this error in the hope that it could be dealt with in connection with a revision of Newcomb's value of general precession. I consider the elimination of a fictitious non-precessional motion of the equinox as a necessity, and the FK5 should be freed from this error independently of a decision on a change in the precessional values. The system of positions and proper motions in right ascension in the FK5 will rest on observations after 1900 only. It would therefore be illogical, if the zero-point determination would contain a relic from the 19th century that has been recognized as a systematic error due to changes in the techniques of observation.

A similar situation exists with the determination of the obliquity ε of the ecliptic. Individual corrections $\Delta\varepsilon$ to Newcomb's obliquity from observations of the Sun and planets show a pronounced secular change of the order of

$$\Delta\dot{\varepsilon} = -0\overset{\prime\prime}{.}3 \text{ per century,}$$

if observations before 1900 are taken into account. This change cannot be explained by planetary precession. The suspicion that we are also here dealing with systematic errors of old observations is supported by the fact that the observations after 1900 do not yield a significant secular change. If there were a real change, I would agree with Aoki (1968) that the equator of the Earth is performing a rotation about an axis passing through the vernal equinox and that the effect would need a physical explanation.

In summarizing this part of my talk, it can be said that the improvement of the equator point and the equinox require an extensive investigation of all relevant observations of the Sun and other bodies of the planetary system and of the consequences of changes for the system of the FK5. Fortunately, this investigation can be made independently from the study of systematic differences between the FK4 and modern observations which will contribute to the elimination of regional errors in the fundamental system.

The derivation of individual corrections to the data in the FK4 will rest upon all observations that have become available from about 1950 onwards and that have been carried out by absolute and differential techniques. The problem of determining weights to the observations will have to be reconsidered in view of the large material which is already incorporated in the FK4. In the past, there has often been some suspicion that new observations, which have been added in the formation of a new fundamental catalogue, have not received the proper weight for updating the fundamental system. Every care has to be taken that suspicion of this sort is unfounded.

5. Compilation of a Supplement to the FK5

Among the observational material now available there will be data for stars which cannot be included in the new system without loss of accuracy. For such stars the computer methods applied to the fundamental stars may, however, yield positions and proper motions which are much better than those which can be derived from photographic plates taken at several epochs. It appears attractive to include such stars in a Supplement to the FK5 whenever it is possible to give their positions at mean epoch with an accuracy appreciably better than $\pm 0\overset{s}{.}01 \sec \delta$ and $\pm 0\overset{''}{.}1$ (standard errors in right ascension and declination, respectively) and their proper motions with an accuracy better than $\pm 0\overset{''}{.}40$ per century in each coordinate. This aim has been reached for the stars in the FK3 Sup, and these have turned out to be exceedingly useful for many purposes. We are therefore planning to compile a Supplement Catalogue to the FK5 resulting partly from a by-product of the work at the FK5. To another part this Supplement may incorporate entirely new material, namely, the positions of faint objects in the system of the FK5.

Among the most attractive faint objects that may be taken into account are the optical counterparts of compact extragalactic radio sources. They are suited for relating the fundamental system to the radio astrometric system of extragalactic sources. Moreover, stars of our galactic system which are radio sources are of interest. For both classes of objects optical positions can be obtained by photographic means with an accuracy of about $\pm 0\overset{s}{.}015 \sec \delta$ and $\pm 0\overset{''}{.}15$ (standard errors in right ascension and declination, respectively). Although this accuracy is inferior to that of radio astrometric measurements, and although the positions of optical counterparts may not be strictly identical with those of the radio sources, the optical astronomers should not be discouraged from proceeding with the photographic measurements. The Supplement to the FK5 may contribute to the encouragement of measurements in the future, and with this purpose in mind, it should be considered as a challenging sample of stars for future work.

Finally, reference shall be made to some papers giving information on results of observations and problems connected with the improvement and extension of FK4. Desiderata for a new fundamental catalogue have been described by Fricke and Gliese (1968), who have also reported on new observations. More recent observations have been discussed by Gliese (1970, 1974). From the reports by Anguita *et al.* (1968, 1971,

1974) and Høg and Nikoloff (1974) it has become clear that the FK4 system in right ascension requires an appreciable correction in the declination zone $-50°$ to $-80°$. Zverev (1968) has reported on absolute observations made in the USSR that indicate some significant regional deviations from the FK4 in right ascension, and Yasuda (1972) has reported on semi-absolute observations which resulted in corrections to right ascensions. Fricke (1970, 1973) has considered the question of whether the fundamental system can be replaced by a reference system of galaxies, and has described the prospects for an extension of the fundamental system to faint objects and radio sources. The present situation as to fundamental systems of positions and proper motions has been reviewed by Fricke (1972).

References

Anguita, C., Carrasco, G., Loyola, P., Shishkina, V. N., and Zverev, M. S.: 1968, in L. Perek (ed.), *Highlights of Astronomy*, D. Reidel Publ. Co., Dordrecht-Holland, p. 292.

Anguita, C., Zverev, M. S., Carrasco, G., and Polozhentsev, D. D.: 1971, *Izv. Glav. Astron. Obs. Pulkovo*, No. 189–190, 83.

Anguita, C., Carrasco, G., Loyola, P., Naumova, A. A., Nemiro, A. A., Polozhentsev, D. D., Shishkina, V. N., Taibo, R., Timashkova, G. M., Varin, M. P., Varina, V. A., and Zverev, M. S.: 1974, this volume p. 35.

Fricke, W.: 1970, in W. J. Luyten (ed.), 'Proper Motions', *IAU Colloq.* **7**, 105.

Fricke, W.: 1972, *Ann. Rev. Astron. Astrophys.* **10**, 101.

Fricke, W.: 1973, *Mem. Soc. Astron. Italiana* **43**, 751.

Fricke, W. and Gliese, W.: 1968, in L. Perek (ed.), *Highlights of Astronomy*, D. Reidel Publ. Co., Dordrecht-Holland, p. 301.

Gliese, W.: 1970, in W. J. Luyten (ed.), 'Proper Motions', *IAU Colloq.* **7**, 146.

Gliese, W.: 1974, this volume p. 31.

Høg, E. and Nikoloff, I.: 1974, this volume, p. 79.

Yasuda, H.: 1972, *Publ. Astron. Soc. Japan* **24**, 247.

Zverev, M. S.: 1968, in L. Perek (ed.), *Highlights of Astronomy*, D. Reidel Publ. Co., Dordrecht-Holland, p. 286.

DISCUSSION

Vasilevskis: If discrepancies between the Lick and Pulkovo proper motions are explained, are you going to make use of proper motions measured with reference to galaxies?

Fricke: We will not use them in the FK5, since there we need to know positions at various epochs which we can compare with the positions provided by observations made with different instruments. The fundamental proper motions will be derived from the positions. We may, however, consider the possibility of taking them into account in the Supplement to the FK5, if positions become known from other sources.

Murray: I don't see why you ignore the information on the motion of the equinox given by the stellar kinematics, since you have to use a kinematic hypothesis to determine the precession.

Fricke: In deriving a fundamental catalogue from observations alone, we have to determine the equinox from absolute observations of the Sun and other bodies of the planetary system, and if the determination yields an unexplained motion of the equinox then we shall be in a difficult position. Boss was faced with that difficulty in 1906, and he proposed to eliminate the unexplained motion of the equinox by making use of the information given by stellar kinematics, because he considered the planetary observations to be of poor quality. At present, we have clear indications that the observations of the Sun and planets from 1900 onwards do not show a secular variation of the equinox. We will investigate these observations in detail, and

I hope they will be good enough to eliminate the motion of the equinox without recourse to stellar kinematics.

Strand: In view of the proposal to provide a Supplement to the FK5 containing positions of radio sources, emphasis should be placed on the need to observe these optically, perhaps by co-operation between several observatories. Such co-operation will probably be necessary since few observatories have instrumentation to cover the entire range from the brighter stars to the optically faint objects associated with the radio sources.

Fricke: I agree entirely with Dr Strand's remark.

Wood: I would be inclined to wager that it would be a quotation from Clemence that coordinates should be defined to conform to an operational method of determining them. This seems to make it difficult to get away from the equator as a fundamental plane.

Fricke: I agree with Clemence's opinion.

Gubbay: Does Prof. Fricke wish to include positions of compact radio sources or only their optical counterparts in the FK5? If the radio positions per se are to be included, how should we develop an ethos for radio position measurement acceptable to the Catalogue? Should we have a table of primary reference sources? Furthermore the ephemeral quality of some compact components and the dependence of their 'visibility' on the orientation of the baseline may introduce difficulties in relating different observations over a period of time.

Fricke: Radio position measurements per se should be included as well as that of their optical counterparts. We look to the radio astronomers to help resolve the problems of cataloguing radio position measurements.

Klock: In regard to Prof. Fricke's concern on the improvement of the equator point with solar observations made in the last twenty years, it is noted that a serious problem exists, inasmuch that the number of transit circles making solar observations has dwindled to an alarmingly small number today.

OBSERVATIONS RELEVANT TO THE IMPROVEMENT OF THE FUNDAMENTAL SYSTEM

W. GLIESE

Astronomisches Rechen-Institut, Heidelberg, F.R. Germany

Abstract. Since 1950 about 150 observational catalogues have become available which may contribute to the improvement and extension of the FK4. Twenty-five of these have been classfied as 'absolute' or 'quasi-absolute' by their authors. Part of the observations, however, are not yet available in publications or in a form suited for data processing. Some of the absolute observations in R.A. suffer from insufficient determination of the azimuth and others in Decl. from unclear elimination of flexure. Main deficiencies in the system of the FK4 are inaccurate positions and proper motions in right ascensions south of $-35°$.

A search for observations relevant to an improvement and extension of the fundamental catalogue yielded about 150 catalogues published after 1950 which have not been included in the FK4. About twenty more catalogues have been observed which are not yet published.

Most of the positions have been observed differentially and will allow an improvement of the individual positions and proper motions.

Only about thirty catalogues will contribute also to an improvement of the fundamental system of positions and proper motions.

Such observations are called 'absolute' even if they do not include the determination of the equator and the equinox. They should fulfil two conditions:

(1) The observed positions have been determined without any reference to stellar positions in a basic catalogue, or, if such basic positions have been used, their systematic effects must have been eliminated by a suitable reduction.

(2) The effects of instrumental errors on the observed positions must be eliminated.

The first condition is often fulfilled only for parts of the observed data. The second condition will never be fulfilled exactly but only approximately.

Different observational methods have been employed and the behaviour of instruments varies from catalogue to catalogue. Therefore, data classified as 'absolute' by their authors will contribute with different weights to a fundamental system but there are no strong rules for computing such weights.

The compilers of a fundamental system need detailed information on methods of observations and on instrumental behaviour during the observational periods. Introductions and texts to observed catalogues sometimes describe the methods of observation and reduction in insufficient detail; in some cases, questions concerning old catalogues will never be answered.

Special attention should be given to the determination of the azimuth in observations of absolute right ascension. In declination, the elimination of flexure errors will influence significantly the weights of absolute observations. Preliminary reviews on possible errors in the FK4 system have been given at Prague (Fricke and Gliese, 1968) and at Minneapolis (Gliese, 1970). The declination system seems to be virtually correct. The large $\Delta\alpha_\delta$ error in southern declinations is the weakest point in the FK4. Dr Anguita will present a detailed report on this situation later in this symposium. I should emphasize that, from the new observational series in the southern hemisphere, a reliable fundamental right ascension system at epoch about 1968 will be derived in a few years.

However, most astronomers are much more interested in a fundamental proper motion system free from errors than in positions accurate just for one special epoch.

It seems to me very difficult to obtain first epoch absolute right ascension catalogues relevant to the derivation of a good proper motion system in the south. Neither FK3 nor N30 nor FK4 have yielded such a system. Their southern systems have been based mainly on the catalogue series observed with Gill's Transit Circle at the Cape. In spite of all careful work of the observers, unexplained variations of the instrumental system occurred during its fifty years of action. Some at least of these catalogues, for example the 1.Cp$_{50}$, should not be used for the compilation of a fundamental right ascension system.

On the other hand, it is very interesting to notice that an intensive but short series observed during the Cape winter 1936 already shows the $\Delta\alpha_\delta$ errors in the fundamental systems. The special value of this small catalogue lies in the very accurate determination of the azimuth by many successive culminations observed from April to September.

We have to look once more very carefully into the work done at the Cape Observatory in the first decades of this century, but I have some doubt whether the observations at Cordoba and at Melbourne can contribute significantly to this problem.

Therefore, although I expect a fundamental system remarkably improved especially in the southern hemisphere in the near future, I am sceptical about any improvement in the right ascension proper motion system south of $-45°$.

References

Fricke, W. and Gliese, W.: 1968, in L. Perek (ed.), *Highlights of Astronomy*, D. Reidel Publ. Co., Dordrecht-Holland, p. 301.

Gliese, W.: 1970, in W. J. Luyten (ed.), 'Proper Motions', *IAU Colloq.* **7**, 146.

DISCUSSION

Stoy: I think that the odd behaviour of the southern FK4 right ascension system is to be traced back to the different methods used for observing the slow moving polar stars which has resulted in an apparent twist of the polar cap relative to the rest of the sky. Once such a twist was in the system it was perpetuated by the practice since the publication of FK3 of adopting the FK3 places for both the clock and azimuth stars. The situation has been further complicated by the use of a different selection of azimuth stars for different catalogues, but, now it is known that this effect is present, it should be possible to trace it through the various catalogues and correct for it.

Gliese: I agree with you that a re-investigation of former Cape catalogues based on improved positions of the azimuth stars will result in corrected catalogue systems but not necessarily in independent 'absolute' systems. In any case, an enormous amount of work would be needed.

REPORT ON THE COMPILATION OF THE
PFKSZ-2 CATALOGUE

Ia. S. IATSKIV and A. N. KURIANOVA

Golossejevo Observatory, U.S.S.R.

and

D. D. POLOZHENTSEV and M. S. ZVEREV

Pulkovo Observatory, U.S.S.R.

The recompilation of PFKSZ (Zverev and Polozhentsev, 1958) catalogue is undertaken in connection with a number of new observations of FKSZ stars which are now available.

This work is being carried out at Pulkovo and Golossejevo Obrvatories. In addition to the catalogues included in PFKSZ (designated PFKSZ-1 in Table I) it is planned to use 3 more groups of catalogues. Table I gives the data for all groups.

TABLE I

Groups of catalogues to be used in compilation of the PFKSZ-2 catalogue

Group	Mean epoch	Number of catalogues		Number of observations of 1 star		$\varepsilon_\alpha \cos\delta$	ε_δ
		α	δ	α	δ		
PFKSZ-1	1949	8	8	29	37	$\pm 0{.}^{s}020$	$\pm 0{.}''49$
KSZ	1959	3	3	11	9	0.030	0.49
AGK3R	1960	12	12	30	30	0.017	0.38
FKSZ-2	1961	7	5	43	27	0.024	0.42
Total (PFKSZ-2)	1957	30	28	113	103	0.025	0.44

In this table PFKSZ-1 combines catalogues used in the compilation of PFKSZ, KSZ consists of catalogues of FKZ stars observed in narrow zones at 7 observatories (FKSZ are included). The FKSZ-2 group contains the FKSZ catalogues completed after publication of the FKSZ.

A comparison of the catalogues with PFKSZ is now being made. The Brosche (1966) method is used. 10 catalogues in R.A. and 9 catalogues in Declination have been compared. It is supposed that PFKSZ-2 will be characterized by the following mean errors:

$$\varepsilon_\alpha \cos\delta = \pm 0{.}^{s}002, \qquad \varepsilon_\delta = \pm 0{.}''04.$$

Gliese, Murray, and Tucker, 'New Problems in Astrometry', 33–34. All Rights Reserved.

The PFKSZ-2 proper motion system will be derived from the catalogues at 3 epochs:

1957 – PFKSZ-2
1930 – AGK2, AGK2A and Yale
1900 – GC and others.

The compilation of PFKSZ-2 can be rationally considered as the first step towards including FKSZ stars into the FK5. Consultation with the Astronomisches Rechen-Institut at Heidelberg in connection with this work is very desirable.

References

Brosche, P.: 1966, *Veröff. Astron. Rechen-Inst. Heidelberg*, Nr. 17.
Zverev, M. S. and Polozhentsev, D. D.: 1958, *Trudy Glav. Astron. Obs. Pulkovo* **72**, 5.

ON THE IMPROVEMENT OF THE α_δ SYSTEM OF THE FK4 CATALOGUE*

C. ANGUITA, G. CARRASCO, and P. LOYOLA

Departamento de Astronomía, Universidad de Chile, Chile

and

A. A. NAUMOVA, A. A. NEMIRO, D. D. POLOZHENTSEV, T. A. POLOZHENTSEVA,
V. N. SHISHKINA, R. TAIBO, G. M. TIMASHKOVA, M. P. VARIN,
V. A. VARINA, and M. S. ZVEREV

Pulkovo Observatory, Leningrad, U.S.S.R.

Abstract. About 45 000 absolute and semi-absolute observations of right ascension of the FK4 stars were obtained during 1963–1972. The observations were made with the three instruments of the Cerro Calán Observatory jointly by Chilean and Soviet astronomers. As a result, systematic $\Delta\alpha_\delta \cos\delta$ errors were found to exist in the FK4 catalogue. For the reduction of star observations according to the SRS program, it is necessary to derive the system of the $\Delta\alpha$ corrections for the FK4.

* The full text of this paper appears in *Inf. Bull. South. Hemisph.*, No. 22, April 1973.

ABSOLUTE DETERMINATION OF RIGHT ASCENSIONS OF STARS AT HIGH GEOGRAPHICAL LATITUDE DURING THE POLAR NIGHT

G. M. PETROV

Nikolajev Observatory, U.S.S.R.

Abstract. It is proposed that transit observations on the islands of Spitsbergen ($\varphi = 78°$) and Ross ($\varphi = -78°$) be organised in order to determine absolute right ascension by observing continually throughout the polar night.

In determinations of absolute right ascensions of celestial bodies the most important tasks are derivations of the absolute instrumental azimuth and smoothing the R.A. of clock stars from the initial catalogue.

A determination of the absolute azimuth (or Bessel's n), according to the Pulkovo method, is made from observations of the same star at two culminations with the time interval of 12 hours, i.e. under quite unequal conditions. Meteorological factors are known to influence on the refraction and also the instrument. Since the precise allowance for these influence has not been achieved as yet, the observational results may be expected to be distorted by a systematic error varying during the year.

The above difficulties could be reduced if observations be made during the whole polar night at the geographical latitude $\varphi \approx 80°$. At this latitude, a night lasts up to 90 days, most of the stars being observable in both culminations and meteorological characteristics being unchanged with the 24-h period. We suppose that under these conditions more precise results could be obtained.

The best place for such observations in the northern hemisphere seems to be Spitsbergen island. The Gulf Stream has a beneficial influence on the climate of the island. For example, in one settlement, Piramida, ($\varphi = +78°$) during last 7 years, there were 28 cloudless days (on the average) within one polar night. The mean value of the air temperature was about $-20.3\,°C$ and the wind speed in clear weather did

TABLE I

Proportion of cloudless days and mean
monthly temperatures at McMurdo
10 years

Month	Percent of clear days	Temperature (°C)
April	17.1%	−21.7
May	28.3	−22.9
June	26.6	−24.1
July	34.7	−26.5
August	30.9	−27.3

not exceed 2 m s^{-1}. In the southern hemisphere the village of McMurdo on the Ross island ($\varphi = -78°$) is apparently also quite a suitable place. The characteristics of this place are given in Table I.

At Piramida and McMurdo, snow cover is absent in summer and an astronomical pavilion can be built on permanent frost. Thus the winds, blowing with the rate of 40 m s^{-1} though very rare, would not destroy it.

In the near future we intend to commence the FK4 star observations with the reversible photoelectric transit instrument on Spitsbergen. According to the program the observations will last during 3 polar nights; the stars will be observed not less than 10 times in each culmination.

ON STARS WITH THE MOST FAVOURABLE BACKGROUND
IN OBSERVATIONAL HISTORY

G. VAN HERK

Leiden Observatory, The Netherlands

The stars in the southern hemisphere with the best observational history in meridian work have been identified and listed at Leiden according to their positions in the sky. Variable and double stars and all stars with $|b| \leqslant 10°$ were omitted. The work could be carried out only for the section from 0^h to 16^h right ascension. The criterion for good observational history was either to have eight or more catalogue entries in the GFH, or to have five or more entries in the Hamburger IdS.

The first group mentioned contains 5109 stars, 4572 of them having some entries in the Hamburger Index as well. In the second group, 698 stars were found, all of them having entries in the GFH. There are 115 stars which meet the criteria for both groups. The numbers of stars occurring in fields 20 minutes wide in α and 0.05 in $\sin \delta$ is shown

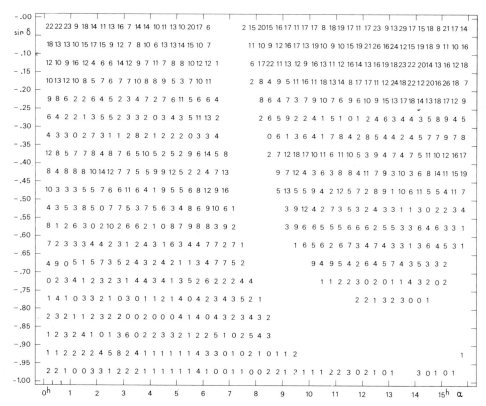

Fig. 1.   Number of stars, south of $\delta = 0°$, zone $|b| > 10°$, (without variables or close doubles) for which 8 or more entries in the GFH or for which 5 or more in the Hamburger Index is found.

Gliese, Murray, and Tucker, 'New Problems in Astrometry', 39–40. All Rights Reserved.

in Figure 1. As is to be expected, most of these stars are situated in a zone within 15° south of the equator.

For the greatest portion of the sky, the distribution of good-history stars is too poor to make use of them in connection with galaxies. It should be remembered that the fundamental stars do not occur in the volumes searched.

We cannot remedy history so we have to proceed as fast as possible with covering the sky by photographic means.

DISCUSSION

Dieckvoss: I am in favour of omitting observations made before 1900, the errors of old meridian catalogues being as they are.

Eichhorn: I think that the question of the value of old catalogues has been oversimplified in the past. The question of whether an old catalogue can be salvaged by determining the systematic errors of the positions in it depends on a number of parameters, among others the homogeneity of the errors and the number of stars in the catalogue in question. It is the accuracy with which the systematic errors can be determined which is most important for utilizing an old catalogue. Brosche's work has been an important step in this direction. The statistical analysis is complicated, and we are now in a position to use a better criterion for using or not using a catalogue than whether its epoch is before or after 1900.

Van Herk: In 1845 a paper was published in which observations up to 1843 had been used to demonstrate the non-rectilinear motion of Sirius. The period of the 'wobble' was predicted to be very close to 50 years (we have it now 49.99) and the amplitude was estimated about 5 to 10% in error of the present value. It was Bessel's paper, and he could not have done it if the observations had been as bad as all that.

SOUTHERN REFERENCE STAR PROGRAM:
PROGRESS REPORT

J. L. SCHOMBERT

U.S. Naval Observatory, Washington, D.C., U.S.A.

The observational programs for the Southern Reference Star Program are rapidly drawing to a close. Observations have been completed, except for a few at San Juan and a small percentage of zone $-52°$ to $-64°$ at the Cape. The Cape zone $-64°$ to $-90°$ has been cancelled. The current state of the observations is shown in Table I.

Final results have now been received from Abbadia, Bordeaux, Tokyo, USNO 6-in. and Cape (zone $-40°$ to $-52°$). These results are in machine readable form and preliminary investigations have begun. Results from several other observatories are expected in a short time.

TABLE I

Status of the SRS programme – 1 June 1973

Observatory	Zone	Commitment	Date started	Completed		Final results received or expected
				Obs'ns	Red'ns	
Abbadia	+ 5° to −15°	1560 × 4	62.3	100%	100%	1968.8
Bordeaux	+ 5 to −15	1560 × 4	62.5	100	100	1970.5
Bucharest	+ 5 to −10	1176 × 4	62.6	100	97	1975.0
Nicolaiev	0 to −20	5984 × 2	64.3	100	98	1974.0
San Fernando	−10 to −30	3709 × 4	63.3	100	100	1973.5
Tokyo	−10 to −30	3560 × 4	63.3	100	100	1972.7
USNO 6-in.	+ 5 to −30	8706 × 2	66.5	100	100	1973.3
		1233 × 4				
Bergedorf (Bickley)	+ 5 to −90	20495 × 4	67.9	100	95	1974.5
Cape	−30 to −40	10082 × 4	61.3	100	97	1974.3
	−40 to −52			100	100	1966.0
	−52 to −64			80	75	1974.5
	−64 to −90			−	−	−
San Juan	−40 to −90	7190 × 2	69.5	99.4	100	1974.3
Santiago −	−25 to −47	5992 × 4	63.1	100	60	1974.0
Pulkovo	−47 to −90	5504 × 4		100	75	1975.0
USNO 7-in.	+ 5 to −20	7683 × 2				
(El Leoncito)	−20 to −75	12121 × 4	68.7	100	71	1975.0
	−75 to −75SP	1382 × 4				

INVESTIGATION OF MAGNITUDE EFFECT IN THE AGK3R CATALOGUE

J. L. SCHOMBERT and T. E. CORBIN

U.S. Naval Observatory, Washington, D.C., U.S.A.

Abstract. The AGK3R Catalogue and the FK4 observations made in connection with this catalogue were examined for magnitude effects. Each observatory's observations were compared to the mean position for all stars. A comparison was also made with the stars grouped by spectral type. The magnitude effects found were very small.

1. Introduction

The AGK3R Catalogue is a catalogue of 21 499 stars between $+90°$ and $-5°$ declination. These stars were observed between 1956 and 1964 with eleven transit instruments at ten different observatories, two instruments being located at the U.S. Naval Observatory. Usually, a star was observed twice with five instruments, although there were many cases where four observations were made at one observatory. The observations were reported to the U.S. Naval Observatory where they were analyzed and each participant's observations reduced to the fundamental system. A final catalogue was then formed. It is these observatory means, for both the AGK3R stars and the FK4 stars, that have been examined for the existence of residual magnitude effects.

2. Comparison of Observatory Means

After each observatory's observations had been reduced to the FK4 system, an average for each star was formed by use of an appropriate weighting factor, and a final catalogue of positions was determined. For each star observed by each instrument, the observed position was reduced to the epoch of the final catalogue using proper motions from the Smithsonian Astrophysical Observatory Catalog. Differences, (Obs. − Cat.), were then formed for all stars, AGK3R and FK4. These differences were solved by a least squares method for a constant displacement and a rate with magnitude using the following equations:

$$\text{Constant displacement} = \overline{\Delta\alpha} = \frac{\sum \Delta\alpha}{N},$$

where

$\Delta\alpha$ = difference between observatory mean and final position, (Obs. − Cat.)

N = number of stars

and

$$\text{Rate/magnitude} = \Delta\alpha_m = \frac{\sum (\Delta\alpha \Delta m)}{\sum (\Delta m)^2},$$

Gliese, Murray, and Tucker, 'New Problems in Astrometry', 43–55. All Rights Reserved.

where

$\Delta m =$ difference between a star's magnitude and the average
magnitude for all stars of that observatory.

Similar equations were used for declination. The results of this solution are shown
in the graphs of Figures 1 through 6 and in Tables I and II.

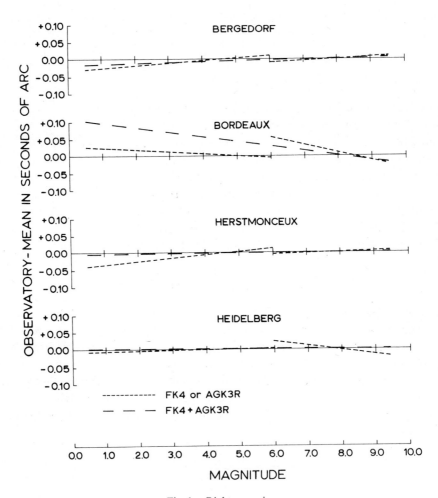

Fig. 1.   Right ascension.

For the most part the slopes are quite small, however, for a few observatories there
seems to be a small magnitude effect. The smaller ones are about the same size as
their mean errors, whereas the larger ones are about 2 to 4 times the mean error. In
the worst case the rate is about 0″.1 over an 8 mag. range.

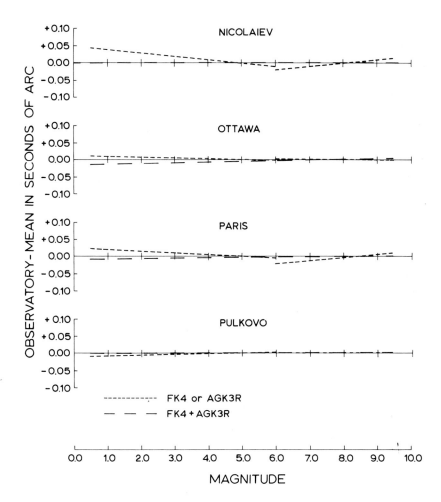

Fig. 2. Right ascension.

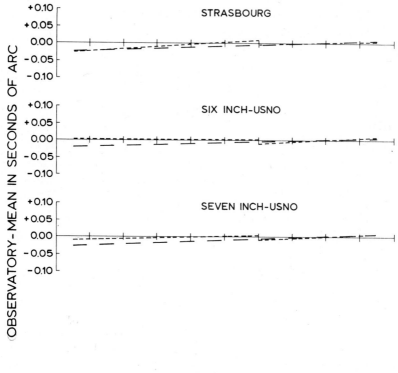

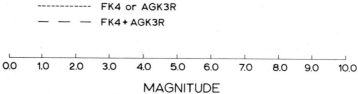

Fig. 3.   Right ascension.

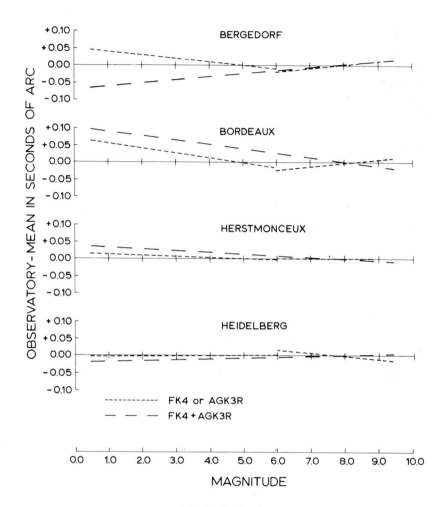

Fig. 4.   Declination.

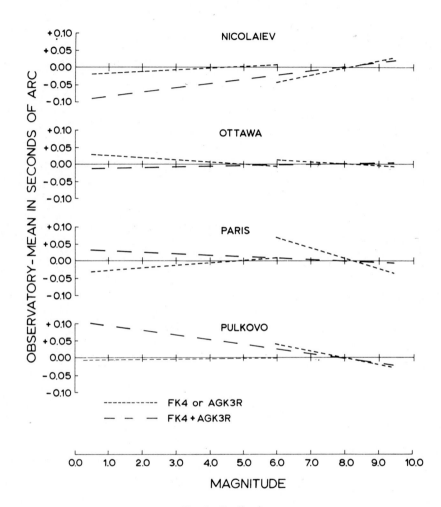

Fig. 5. Declination.

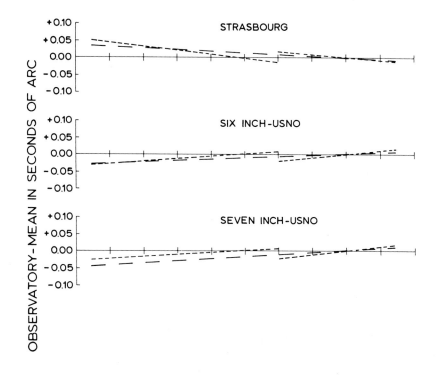

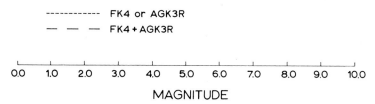

Fig. 6.   Declination.

TABLE I

Observatory – Catalogue

Observatory	AGK3R+FK4		AGK3R		FK4	
	$\overline{\Delta\alpha}$	$\Delta\alpha_m$	$\overline{\Delta\alpha}$	$\Delta\alpha_m$	$\overline{\Delta\alpha}$	$\Delta\alpha_m$
Bergedorf	−0″.0009	+0″.0027	−0″.0011	+0″.0057	+0″.0002	+0″.0070
Bordeaux	−0.0335	−0.0138	−0.0374	−0.0225	+0.0044	−0.0060
Herstmonceux	−0.0056	+0.0015	−0.0057	+0.0033	−0.0015	+0.0108
Heidelberg	+0.0105	−0.0004	+0.0120	−0.0126	−0.0039	−0.0006
Nicolaiev	−0.0074	+0.0018	−0.0080	+0.0102	+0.0045	−0.0057
Ottawa	+0.0111	+0.0022	+0.0134	−0.0009	+0.0018	−0.0036
Paris	−0.0006	+0.0012	−0.0009	+0.0096	+0.0023	−0.0051
Pulkovo	+0.0015	+0.0004	+0.0017	+0.0003	+0.0008	+0.0022
Strasbourg	+0.0015	+0.0034	+0.0024	+0.0028	−0.0087	+0.0063
USNO 6-in.	+0.0068	+0.0027	+0.0074	+0.0057	+0.0015	0.0000
USNO 7-in.	+0.0113	+0.0039	+0.0120	+0.0050	+0.0009	+0.0024

TABLE II

Observatory – Catalogue

Observatory	AGK3R+FK4		AGK3R		FK4	
	$\overline{\Delta\delta}$	$\Delta\delta_m$	$\overline{\Delta\delta}$	$\Delta\delta_m$	$\overline{\Delta\delta}$	$\Delta\delta_m$
Bergedorf	+0″.0293	+0″.0089	+0″.0316	+0″.0104	−0″.0018	−0″.0102
Bordeaux	−0.0666	−0.0124	−0.0735	+0.0103	+0.0006	−0.0151
Herstmonceux	−0.0282	−0.0042	−0.0296	+0.0006	−0.0064	−0.0039
Heidelberg	+0.0232	+0.0022	+0.0252	−0.0086	+0.0031	+0.0018
Nicolaiev	+0.0251	+0.0128	+0.0264	+0.0208	+0.0013	+0.0054
Ottawa	+0.0132	+0.0023	+0.0163	−0.0056	+0.0006	−0.0070
Paris	+0.0153	−0.0041	+0.0166	−0.0296	+0.0039	+0.0073
Pulkovo	−0.0305	−0.0121	−0.0315	−0.0193	−0.0103	+0.0016
Strasbourg	+0.0058	−0.0049	+0.0052	−0.0083	+0.0117	−0.0118
USNO 6-in.	+0.0031	+0.0038	+0.0033	+0.0100	+0.0014	+0.0060
USNO 7-in.	+0.0090	+0.0060	+0.0097	+0.0114	+0.0004	+0.0057

3. Comparison by Spectral Type

The differences were also grouped by spectral type. K, M, and R stars formed the red group and O, B, A, F and G the blue. The same type solution was made for these groups and the results are given in Tables III and IV. As can be seen, there seems to be no significant relationship with color.

4. Comparison with the FK4

A comparison was made between the FK4 itself and the observational means of the

TABLE III

Observatory – Catalogue

Observatory	Red stars						Blue stars					
	AGK3R+FK4		AGK3R		FK4		AGK3R+FK4		AGK3R		FK4	
	$\overline{\Delta\alpha}$	$\Delta\alpha_m$	$\overline{\Delta\alpha}$	$\Delta\alpha_m$	$\overline{\Delta\alpha}$	$\Delta\alpha_m$	$\overline{\Delta\alpha}$	$\Delta\alpha_m$	$\overline{\Delta\alpha}$	$\Delta\alpha_m$	$\overline{\Delta\alpha}$	$\Delta\alpha_m$
Bergedorf	−0″.0002	−0″.0002	−0″.0006	+0″.0008	+0″.0050	+0″.0126	−0″.0071	+0″.0018	−0″.0069	+0″.0038	−0″.0078	+0″.0042
Bordeaux	−0.0191	−0.0134	−0.0215	−0.0244	+0.0020	+0.0028	−0.0542	−0.0141	−0.0612	−0.0220	+0.0093	−0.0104
Herstmonceux	−0.0060	+0.0003	−0.0065	+0.0028	+0.0003	+0.0110	−0.0050	+0.0022	−0.0050	+0.0034	−0.0053	+0.0108
Heidelberg	+0.0071	+0.0008	+0.0080	−0.0074	−0.0018	−0.0002	+0.0156	−0.0015	+0.0174	−0.0165	−0.0041	−0.0006
Nicolaiev	−0.0018	−0.0034	−0.0020	−0.0048	+0.0014	−0.0051	−0.0234	+0.0010	−0.0252	+0.0136	−0.0087	−0.0050
Ottawa	+0.0111	+0.0039	+0.0131	+0.0014	+0.0011	−0.0070	+0.0110	+0.0010	+0.0141	−0.0086	+0.0027	−0.0024
Paris	−0.0002	+0.0003	−0.0002	+0.0050	+0.0002	−0.0117	−0.0024	+0.0016	−0.0035	+0.0129	+0.0050	−0.0012
Pulkovo	+0.0018	+0.0027	+0.0021	+0.0020	−0.0054	+0.0008	+0.0020	−0.0008	+0.0015	−0.0003	+0.0096	+0.0034
Strasbourg	−0.0030	+0.0042	−0.0032	+0.0066	−0.0015	+0.0267	+0.0090	+0.0036	+0.0119	+0.0010	−0.0188	−0.0027
USNO 6-in.	+0.0027	+0.0013	+0.0029	+0.0022	+0.0012	+0.0002	+0.0138	+0.0042	+0.0150	+0.0096	+0.0011	+0.0002
USNO 7-in.	+0.0084	+0.0058	+0.0092	+0.0068	−0.0017	+0.0002	+0.0156	+0.0026	+0.0167	+0.0026	+0.0036	+0.0032

TABLE IV

Observatory – Catalogue

Observatory	Red stars						Blue stars					
	AGK3R+FK4		AGK3R		FK4		AGK3R+FK4		AGK3R		FK4	
	$\overline{\Delta\delta}$	$\Delta\delta_m$	$\overline{\Delta\delta}$	$\Delta\delta_m$	$\overline{\Delta\delta}$	$\Delta\delta_m$	$\overline{\Delta\delta}$	$\Delta\delta_m$	$\overline{\Delta\delta}$	$\Delta\delta_m$	$\overline{\Delta\delta}$	$\Delta\delta_m$
Bergedorf	+0″0166	+0″0092	+0″0183	+0″0082	−0″0051	−0″0088	+0″0455	+0″0072	+0″0485	+0″0097	+0″0043	−0″0105
Bordeaux	−0.0439	−0.0096	−0.0483	+0.0230	−0.0043	−0.0248	−0.1082	−0.0174	−0.1215	−0.0047	+0.0108	−0.0084
Herstmonceux	−0.0225	−0.0084	−0.0236	−0.0063	−0.0063	−0.0083	−0.0363	−0.0012	−0.0380	−0.0064	−0.0082	−0.0007
Heidelberg	+0.0165	+0.0008	+0.0172	−0.0071	+0.0100	+0.0078	+0.0328	+0.0028	+0.0367	−0.0106	−0.0091	−0.0038
Nicolaiev	+0.0099	+0.0116	+0.0106	+0.0178	−0.0023	+0.0023	+0.0458	+0.0137	+0.0483	+0.0228	+0.0029	+0.0092
Ottawa	+0.0114	+0.0040	+0.0140	−0.0064	−0.0018	−0.0071	+0.0182	+0.0013	+0.0234	−0.0097	+0.0044	−0.0082
Paris	+0.0054	−0.0015	+0.0058	−0.0131	+0.0024	+0.0141	+0.0343	−0.0045	+0.0386	−0.0409	+0.0037	+0.0022
Pulkovo	−0.0207	−0.0169	−0.0217	−0.0231	+0.0004	+0.0165	−0.0464	−0.0094	−0.0473	−0.0176	−0.0300	−0.0069
Strasbourg	+0.0033	−0.0043	+0.0031	−0.0060	+0.0063	−0.0106	+0.0101	−0.0049	+0.0090	−0.0087	+0.0214	−0.0115
USNO 6-in.	+0.0047	+0.0035	+0.0047	+0.0099	+0.0057	+0.0044	−0.0008	+0.0040	−0.0003	+0.0097	−0.0064	+0.0079
USNO 7-in.	+0.0095	+0.0064	+0.0101	+0.0086	+0.0009	+0.0039	+0.0084	+0.0054	+0.0090	+0.0141	+0.0012	+0.0057

FK4 stars, using FK4 proper motions. The results are shown in Table V and Figures 7 and 8. In right ascension, although small, the rate is 3.5 times its mean error, while in declination the rate is insignificant.

TABLE V

FK4 and proper motion system-catalogue

	$\overline{\Delta\alpha}$	$\Delta\alpha_m$	$\overline{\Delta\delta}$	$\Delta\delta_m$	N
FK4 – Cat.	$+0\overset{\prime\prime}{.}0005$	$-0\overset{\prime\prime}{.}0052$	$+0\overset{\prime\prime}{.}0003$	$+0\overset{\prime\prime}{.}0009$	886
PMS – Cat(FK4)	-0.0039	-0.0026	-0.0025	$+0.0005$	870
PMS – Cat(AGK3R)	$+0.0016$	$+0.0052$	$+0.0061$	$+0.0040$	5394
PMS – Cat(AGK3R + FK4)	$+0.0009$	$+0.0024$	$+0.0049$	$+0.0030$	6264
By average magnitude groups:					
PMS – Cat(AGK3R + FK4)	$+0.0062$	-0.0009	-0.0023	$+0.0008$	6264

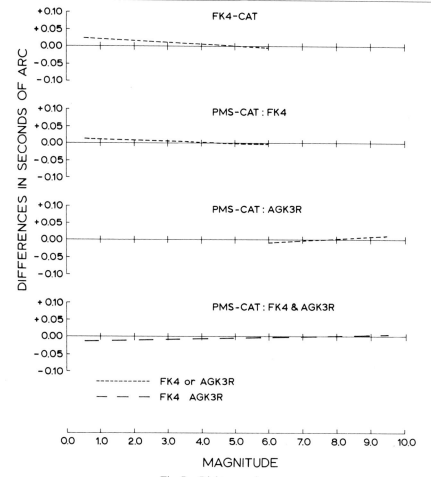

Fig. 7.   Right ascension.

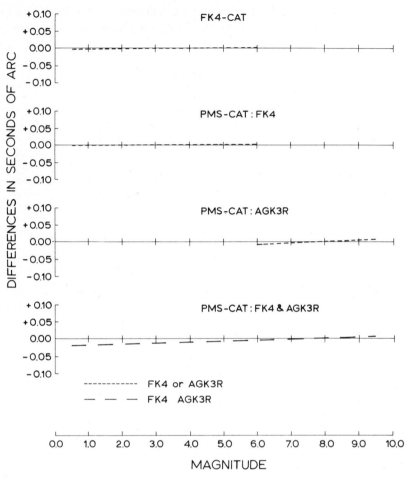

Fig. 8.   Declination.

5. Comparison with the AGK3R Proper Motion System

In deriving proper motions for the AGK3R stars, a basic proper motion system (PMS) is being developed that will be used to bring in more and more of the older catalogues. Carefully selected catalogues that are relatively free of magnitude effect and that had been observed in such a way as to avoid other problems now known to exist in 19th century catalogues were used to form this base system of some 5400 reference stars. In order to compare the AGK3R Catalogue with some other standard, differences were formed between the PMS and the AGK3R at the epoch of the AGK3R observations. This was done for 3 groups: FK4 stars, AGK3R stars and both of these groups combined. While it is realized that the PMS includes the AGK3R, it also includes nine other catalogues ranging in epoch from 1898 to 1958. The mean epoch of the PMS is about 1935. The results of this comparison are shown in Table V and

Figures 7 and 8. In declination, the rate generated by the comparison of the observed FK4 stars with the PMS is near zero and, in right ascension, it is about half that of the comparison with the FK4 itself. Whether this is due to a magnitude effect in the FK4 is uncertain, since, as noted above, the AGK3R is one of the constituents of the PMS.

6. Comparison by Average Magnitude Groups

All of the above analyses have been made by star, that is, each star received equal weight, regardless of its magnitude. This means that in the analyses of the combined FK4 and AGK3R stars, the fainter magnitudes entered into the solution with considerably greater weight than the brighter magnitudes, due to the distribution of stars. An analysis was also made, collecting the differences for each half magnitude for (PMS − AGK3R + FK4). A solution was made giving each of these groups equal weight. The results are shown in Table V. There is little evidence of a significant magnitude effect.

7. Conclusion

There is little evidence of a significant magnitude effect in the AGK3R. If any, it is less than $0\overset{''}{.}01$ mag^{-1} in either coordinate.

DISCUSSION

Van Herk: What magnitudes have been used? The nominal values or the apparent magnitudes as they were during the observations? If the first, you are bound to get too small values for the 'magnitude' effect, perhaps up to 100%.

Klock: I don't know which values were used. The authors of the paper would have to be consulted before answering the question.

Eichhorn: In my judgement, it would be necessary to include in the meridian catalogues and photographic catalogues more early type (or blue) stars than there are now. There is no doubt that the existing star catalogues are afflicted by systematic errors which depend on the color index of the stars in them. These errors can be reliably determined only if there are enough early type stars in the catalogue; otherwise these errors may be difficult to separate from random errors.

Since it is of utmost importance, in questions of galactic kinematics, that there are no color dependent systematic errors in the proper-motion material, I believe it would be advantageous if the compilers and observers of star catalogues would make a conscious effort to include more blue and white stars than there are now, so that it will become possible to determine reliably the color-dependent systematic errors also for the stars at the blue end of the spectrum.

Klock: Schombert and Corbin investigated the possibility of a magnitude or color effect and their results are given in Table III and IV of their paper. Their conclusion was that there was no significant relationship with color.

In regard to the inclusion of more early type stars in meridian circle programs it should be mentioned that the last two Washington 6-in. observing programs included the Blaauw stars, giving about 15–20% early type stars in their list.

MOUVEMENTS PROPRES DES ETOILES DE REPERE DU CATALOGUE PHOTOGRAPHIQUE

R. BOUIGUE

Université de Toulouse, France

Résumé. On examine le problème de l'obtention des mouvements propres à partir des observations meridiennes effectuées dans différents Observatoires, en vue de la détermination de la précision que l'on peut attendre de ces résultats.

On montre, en outre, que ces écarts ne sont pas imputables à l'hétérogénéité des observations et que l'on ne pourra donc les réduire par la simple création d'un seul et même système d'observation.

Gliese, Murray, and Tucker, 'New Problems in Astrometry', 57. All Rights Reserved.
Copyright © 1974 by the IAU.

DISCUSSION AFTER SESSION A

Murray: A fundamental catalogue can contain many sorts of stars for many purposes. It should include sufficient stars at an optimum magnitude for calibration of the overlapping plate surveys. I would suggest that perhaps 3000 stars in a very narrow magnitude range would be sufficient.

Fricke: I hope that we will be able to provide at least 2000 stars in the FK5 within the magnitude range 7.5 to 9 visual, and there will be others fainter than visual magnitude 7.5 in the FK5 Supplement. I am, however, not able at present to make an estimate of the total number of objects fainter than 7.5 in the FK5 Sup.

Lacroute: It is important that Heidelberg provides a large number of fundamental and supplementary stars for the photographic reduction even if the precision is not as good as for the fundamental stars brighter than visual magnitude 7.5.

Bok: (1) Obviously we must urge radio astronomers to do their utmost to add to the list of true radio stars, such as β Persei. Such stars – once known – should be observed assiduously and be included in future fundamental catalogues.

(2) We have heard much about the importance of precise radio positions for future fundamental positions. There is a risk that from the start we shall neglect once again the southern hemisphere. We must do everything in our power to urge radio astronomers to assure that from the start there will be a proper balance between objects at northern declinations and those in the deep south.

Fricke: I suggest that the Resolutions Committee of this Symposium takes into account Dr Bok's remarks in drafting two resolutions.

Robertson (D.S.): In reply to Dr Bok's urging southern hemisphere positional radio astronomy, his words are music to my ears. The three radio telescopes that have been used for this work in the southern hemisphere are Tidbinbilla near Canberra, Island Lagoon, near Woomera and Johannesburg. Island Lagoon has been closed and offered to Australian radio astronomers by NASA. Australian radio astronomers do not appear to want it and it is likely to be demolished. Next year Johannesburg is to be closed and further work will not be possible.

Elsmore: Concerning the choice of radio sources for inclusion in a fundamental catalogue, a working party has been set up in Commission 40 to provide a list of calibration sources for positional work; this list will include sources for which there are no optical counterparts. Co-operation with Commission 40 is therefore advised.

Fricke: Such co-operation would be much welcomed by Commission 8 and, in particular, by my colleagues at Heidelberg and myself.

Dieckvoss: It may be mentioned that at Bergedorf for the declination zones north of $+40°$ the data of the astrographic catalogue were combined with AGK2 and AGK3 to give proper motions with mean errors of $\pm0\overset{''}{.}4$ per century.

CURRENT AND FUTURE PROJECTS FOR SOUTHERN HEMISPHERE REFERENCE SYSTEMS

OBSERVATIONS IN CHILE AND RESULTS RELATED TO SOUTHERN HEMISPHERE SYSTEMS

(Invited Paper)

C. ANGUITA

Observatorio Astronómico Cerro Calán, Departamento de Astronomía, Universidad de Chile, Chile

Abstract. The astrometric observations made at the Cerro Calán observatory with several instruments since 1962 are outlined. The resulting catalogues are in press or approaching publication, and are described in some detail. An account is given of the observational projects currently in progress.

The various catalogues are combined in order to demonstrate the large and well-determined systematic differences in $\Delta\alpha_\delta \cos\delta$ between the Cerro Calán observations and the FK4. It is concluded that the FK4 system requires some improvement if it is to fulfil the requirements of a fundamental reference system in the southern hemisphere.

1. Introduction

Several programmes of observations related to southern hemisphere reference systems have been observed at the Observatorio Astronómico Nacional de Chile since 1962, the year when the observatory was moved to its present location at Cerro Calán, and when a cooperation between Chilean and Soviet astronomers began under an agreement between the USSR Academy of Sciences and the University of Chile. This joint effort includes observations of the SRS and BS programmes with the Repsold Meridian Circle, absolute and quasi-absolute observations of FK4, FK4 Sup and FKSZ stars with the photographic vertical circle, the Zeiss broken transit instrument and the Pulkovo large transit instrument. The last three instruments were sent to Chile by Pulkovo Observatory and installed at Cerro Calán.

A Danjon impersonal astrolabe has been in operation at Cerro Calán since 1965 for determining latitude and time and corrections to right ascensions and declinations of bright fundamental stars. This instrument was installed at the observatory under an agreement between the European Southern Observatory and the University of Chile.

The purpose of this paper is to present the status of the catalogues already compiled from observations made at Cerro Calán Observatory, the progress of the current programmes of observations and finally, a system of corrections $\Delta\alpha_\delta \cos\delta$ to the FK4 positions compiled on the basis of the four catalogues presented and absolute observations of right ascensions made with the Pulkovo large transit instrument.

2. Catalogues Compiled from Observations made at Cerro Calán Observatory

Three catalogues of right ascension: the Santiago-Pulkovo Fundamental 1 (SPF-1), the Santiago-Pulkovo 2 (SP-2) and the Santiago-Pulkovo Fundamental 3 (SPF-3) and a catalogue of right ascensions and declinations, the First Astrolabe Catalogue of

Gliese, Murray, and Tucker, 'New Problems in Astrometry', 63–72. All Rights Reserved.

Santiago have been compiled. The SPF-1, the SPF-3 Catalogues and the First Astrolabe Catalogue are already in press in *Publicaciones del Departamento de Astronomía de la Universidad de Chile* and the SP-2 Catalogue is now being prepared for publication.

The main characteristics of the catalogues are described below.

(1) The SPF-1 Catalogue gives the position of 1044 FK4 fundamental stars with an average mean epoch of observation of 1965.0. The methods of observation and reduction were described by Anguita *et al.* (1971). The catalogue contains the results of observations of Küstner series of the FK4 stars with the Repsold Meridian Circle ($d = 19$ cm, $f = 244$ cm) in the declination zone $+41°$ to $-90°$ (from $-68°$ to $-90°$ in lower culmination). 211 series of 6793 star observations were made from 1963 to 1968. The mean error of one observation reduced to the equator is $\pm 0°020$ for upper culmination and $\pm 0°022$ for lower culmination.

The reductions were made using the quasi-absolute method proposed by Zverev (1969) which implies that the reductions were made in the instrumental system with the conditions that the system coincides with the FK4 fundamental system in the equator and the pole. Figure 1 gives the systematic differences $\Delta\alpha_\delta \cos\delta$ in the sense SPF1 − FK4, and shows the conditions already mentioned. The results of the 82 stars observed in lower culmination were not combined with the results in upper culmination because the instrumental system may show some irregularities that may affect the individual star positions and for this reason they are listed separately at the end of the catalogue.

(2) The SP-2 Catalogue contains the results of observations of FK4 stars in zone observations of SRS stars with the Repsold Meridian Circle. 15160 FK4 star observations were made from 1963 to 1968 in the zone $-47°$ to $-90°$, the Russian

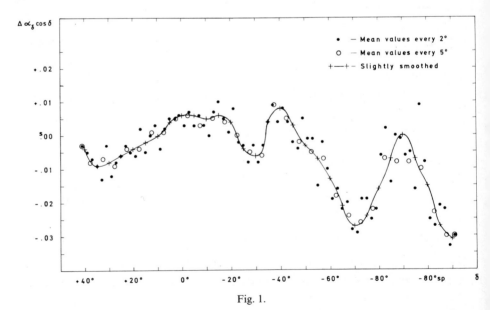

Fig. 1.

commitment of the SRS programme. The mean error of one observation, reduced to the equator is ±0.016 for upper culmination and ±0.020 for lower culmination. The results are just complete in the zone −25° to −47°, the Chilean commitment of the SRS. The mean error of one observation in this zone is ±0.015. All the reductions were made in the FK4 system and in the instrumental system of the Repsold Meridian Circle.

(3) The SPF-3 Catalogue was derived from quasi-absolute observations made with the Zeiss broken transit instrument ($d = 10$ cm, $f = 100$ cm) of a special programme described by Loyola and Shishkina (1968). The programme consisted of 726 stars from

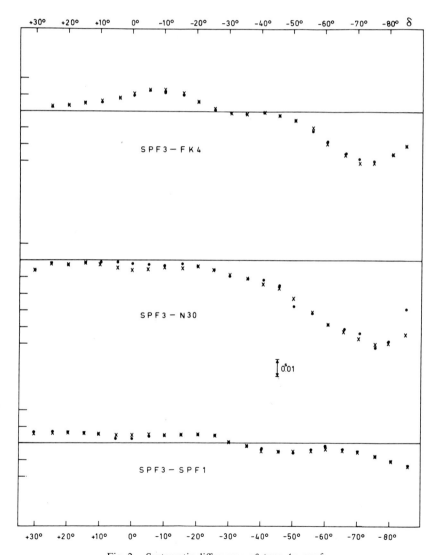

Fig. 2. Systematic differences of type $\Delta\alpha_\delta \cos\delta$.

+40° to −90° including 666 FK4 stars; 42 stars with declination from −75° to −90° were observed in lower culmination. Approximately 12 800 star observations were made from April 1963 to September 1964 and the mean error of one observation is ±0ˢ024.

Besides the series of observations of the main programme, a special programme was regularly observed, once a week, in order to determine the azimuth of the instrument. The series of observations of this auxiliary programme enabled us to smooth

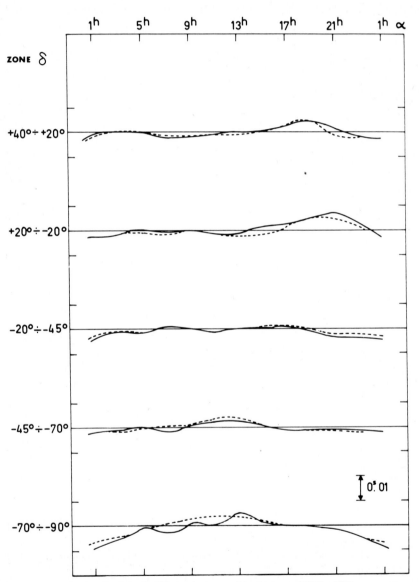

Fig. 3.   Systematic differences $\Delta\alpha_\alpha \cos\delta$ in the sense SPF3 − FK4.

the right ascension system obtained for the zenith zone $(-21°$ to $-45°)$ and to determine the systematic errors of the FK4 in this zone. The SPF-3 Catalogue is tied to the FK4 system in the zenith zone of Santiago but it is free from its $\Delta\alpha_\alpha$ and $\Delta\alpha_\delta$ errors.

Figure 2 shows the systematic differences $\Delta\alpha_\delta \cos\delta$ of the SPF-3 from the FK4, N30 and SPF-1 Catalogues obtained by the classical method of comparing catalogues and the method developed by Brosche (1966). Figure 3 shows that the $\Delta\alpha_\alpha$ errors of the FK4 are smaller than the systematic errors of the type $\Delta\alpha_\delta$.

(4) The First Astrolabe Catalogue of Santiago gives the positions of 540 FK4 and FK4 Sup stars with an average mean epoch of observation of 1969.4. The observing programme was described by Noël (1968). All suitable FK4 and FK4 Sup stars from $-5°$ to $-62°$ and brighter than magnitude 6.3 have been observed. A selection of 233 FK4 and 24 FK4 Sup stars of the zone was arranged in 11 groups of 28 transits distributed equally in azimuth. These are the 'standard groups' which define the system for the Astrolabe Catalogue of Santiago. The remaining stars were observed in the 'catalogue groups' such that each star could be observed in both transits. The first series of 11 catalogue groups of about 300 stars was observed in 1967 and 1968, and the second series of about 200 stars in 1969 and 1970. In 1971 and 1972 the observations of some catalogue groups of both series which had a rather low number of observations were completed. Table I gives the differences in right ascension and declination of the Astrolabe Catalogue and the FK4 as a function of declination. The results in declination seem to indicate that the FK4 declination system is probably more reliable than the right ascension system.

TABLE I

First Astrolabe Catalogue of Santiago
Systematic errors of FK4 fundamental system, as a function of declination
1969.4
Astrolabe-FK4

δ	$\Delta\alpha$	m.e.	N	δ	$\Delta\delta$	m.e.	N
$- 9°\!.5$	$0°\!.001$	$0°\!.001$	63	$- 9°\!.5$	$0''\!.05$	$0''\!.02$	63
-14.0	0.006	0.001	71	-14.0	0.05	0.02	71
-19.7	0.004	0.001	57	-19.7	0.00	0.02	57
-25.2	-0.002	0.002	51	-24.7	-0.05	0.04	46
-30.9	-0.004	0.002	54	-27.6	-0.11	0.07	17
-35.8	0.002	0.003	52				
-41.5	0.002	0.003	52				
-46.4	-0.002	0.003	55	-50.3	-0.03	0.07	14
-52.1	-0.009	0.004	50	-52.8	-0.02	0.04	43
-57.5	-0.023	0.005	46	-57.5	-0.04	0.03	46

3. Current Projects for Southern Hemisphere Reference System

The following projects are at present in progress at Cerro Calán:

(1) SRS Programme. The observations of the Chilean commitment in the zone −25° to −47° (5992 SRS and 805 BS stars) have been carried out with the Repsold Meridian Circle, and the reduction of observations in right ascension has been completed. The final catalogue will be ready by 1975.0.

(2) Absolute right ascension determinations of FK4 and FKSZ stars. This programme of absolute determinations of right ascensions contains 1894 stars including all the FK4 stars in the zone +40° to −90° and all the FKSZ stars from +30° to −90°. All the FK4 stars in zone −70° to −90° and all the FKSZ stars from −75° to −90° are observed in both culminations (130 stars).

40900 individual star observations were made with the Pulkovo large transit instrument ($d = 18$ cm, $f = 240$ cm) from May 1969 to October 1970. Then the objective glass and the ocular were interchanged and a new series of observations started in December 1970. The programme will be continued until the end of this year when there will be the same number of observations as were made in the first position of the instrument.

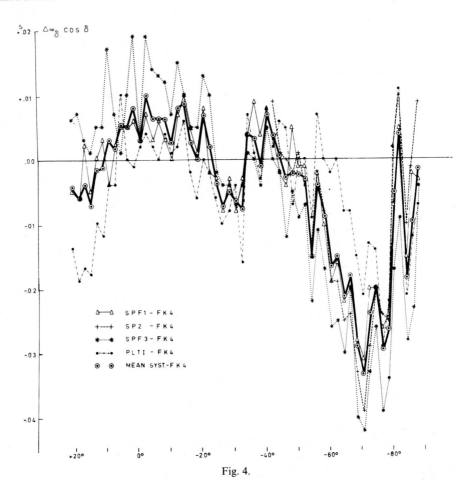

Fig. 4.

(3) FKSZ programme. For inclusion of stars down to 9th magnitude, especially FKSZ stars, into the fundamental system, right ascension and declinations of 651 FKSZ stars in the zone $+40°$ to $-90°$ are being observed with the Repsold Meridian Circle. It is intended to make six observations of each star. The programme started early this year and will take about two years.

(4) Positions of optical counterparts of radio sources. A list of twenty radio sources used as calibrators is being observed with the Double Meniscus Maksutov telescope ($d = 70$ cm, $f/3.0$). The aim of the programme is to give the positions of the optical counterparts of the radio sources in the FK4 system.

4. Compilation of a System of Corrections $\Delta\alpha_\delta \cos\delta$ to the FK4

Using the data of the SPF-1, SPF-3, the southern part of the SP-2 (from $-42°$ to $-90°$) and the series of absolute determinations of right ascension made with the Pulkovo large transit instrument, Anguita *et al.* (1973) compiled a system of corrections $\Delta\alpha_\delta \cos\delta$ to the FK4 system. Figure 4 shows the systematic differences of these catalogues and the FK4, as given in the paper just mentioned. Table II gives the same

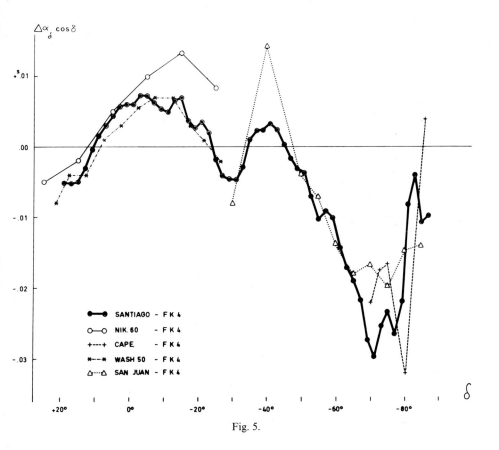

Fig. 5.

TABLE II
Systematic differences $\Delta\alpha_\delta \cos\delta$ (Catalogue – FK4)
(Unit 0.^{s}001)

δ	SPF-1	SP-2	SPF-3	ASTR	PLT1	Mean system	Smoothed mean system	ε_i
+21	− 5		+ 6		−14	− 4.3		±5
+19	− 6		+ 7		−19	− 6.0	− 5.1	6
+17	+ 2		+ 3		−17	− 4.0	− 5.3	5
+15	− 5		+ 1		−18	− 7.3	− 5.0	5
+13	0		+ 5		−10	− 1.7	− 3.0	4
+11	+ 3		+ 5		−12	− 1.3	− 0.4	4
+ 9	− 4		+17		− 4	+ 3.0	+ 1.6	5
+ 7	+ 2		+ 7		− 4	+ 1.7	+ 3.0	3
+ 5	+ 5		+ 1		+10	+ 5.3	+ 4.4	2
+ 3	+ 5		+10		0	+ 5.0	+ 5.8	2
+ 1	+ 6		+19		− 1	+ 8.0	+ 6.0	5
− 1	+ 3		+ 4		+ 2	+ 3.0	+ 6.0	0
− 3	+ 7		+19		+ 4	+10.0	+ 7.3	4
− 5	+ 3		+14		+ 2	+ 6.3	+ 7.2	3
− 7	+ 6		+13		0	+ 6.3	+ 6.3	3
− 9	+ 3		+12		+ 4	+ 6.3	+ 5.4	2
−11	0		+ 7	+ 2	+ 1	+ 2.5	+ 4.6	1
−13	+ 7		+15	+ 5	+ 2	+ 7.2	+ 6.2	2
−15	+10		+10	+ 6	+ 6	+ 8.0	+ 6.6	1
−17	+ 5		+ 5	+ 5	− 2	+ 3.2	+ 3.8	1
−19	+ 1		+ 5	+ 4	− 6	+ 1.0	+ 2.8	2
−21	+ 8	+ 6	+13	+ 2	0	+ 5.8	+ 3.7	2
−23	− 2	+ 5	+10	0	− 2	+ 2.2	+ 2.0	2
−25	− 3	+ 1	− 2	− 2	− 6	− 2.4	− 1.8	1
−27	− 8	0	− 4	− 2	−10	− 4.8	− 4.0	2
−29	− 3	− 2	− 4	− 3	− 8	− 4.0	− 4.4	1
−31	− 8	0	− 7	− 4	− 4	− 4.6	− 4.6	1
−33	− 3	0	− 4	− 2	−16	− 5.0	− 2.8	2
−35	+ 4	+ 3	+ 1	+ 2	+ 7	+ 3.4	+ 1.1	1
−37	+ 9	+ 1	0	+ 2	+ 1	+ 2.6	+ 2.4	1
−39	+ 4	0	− 4	+ 2	+ 3	+ 1.0	+ 2.4	1
−41	+ 8	+ 3	+ 5	+ 2	+ 8	+ 5.2	+ 3.4	1
−43	+ 4	+ 3	0	+ 1	+ 2	+ 2.0	+ 2.5	1
−45	− 2	+ 3	− 2	− 1	+ 6	+ 0.8	+ 0.4	1
−47	− 4	+ 4	−12	− 2	+ 5	− 1.8	− 1.4	3
−49	+ 5	− 4	− 5	− 3	− 7	− 2.8	− 2.6	2
−51	− 1	+ 1	− 9	− 5	0	− 2.8	− 3.0	2
−53	− 1	− 4	− 7	− 7	0	− 3.8	− 6.0	1
−55	−15	−14	−22	∟ 9	− 9	−13.8	− 8.8	2
−57	− 2	− 2	−11	−12	+ 7	− 4.0	− 7.6	3
−59	−10	− 8	−17		0	− 8.8	− 9.5	3
−61	−19	−19	−26		− 2	−16.5	−14.2	4
−63	−16	−19	−25		0	−15.0	−17.0	5
−65	−22	−25	−30		− 8	−21.3	−18.9	4
−67	−20	−24	−20		− 8	−18.0	−21.6	3
−69	−28	−33	−40		−15	−29.0	−27.3	5
−71	−31	−39	−42		−21	−33.3	−29.7	4
−73	−20	−29	−33		−13	−23.8	−25.2	4
−75	−20	−20	−26		−14	−20.0	−23.3	2
−77	−26	−24	−39		−29	−29.5	−26.4	3
−79	−24	−25	−34		−22	−26.3	−21.8	2
−81	− 7	+ 2	−17		+ 2	− 5.0	− 8.0	3
−83	+ 5	+10	− 9		+11	+ 4.3	− 3.8	4
−85	−15	−10	−28		−21	−18.5	−10.6	3
−87	− 2	− 1	−23		−12	− 9.5	− 9.7	3
−89	− 3	+ 9	− 4		− 7	− 1.3		3

differences but adding now the data of the First Astrolabe Catalogue and the northern part of the SP-2 Catalogue ($-20°$ to $-45°$). Column 8 of the table gives the smoothed mean system of corrections to the FK4 and can be seen also in Figure 5 together with the results obtained at the other observatories. The averaged standard error of one correction is $\pm 0\overset{s}{.}0027$.

5. Conclusion

The catalogues compiled from observations made at Cerro Calán have indicated that the FK4 system of right ascension has very large systematic errors of the type $\Delta\alpha_\delta$ and that the system should be improved in order to fulfill the requirements of a fundamental reference system.

Also there are many southern FK4 stars which need large individual corrections in right ascension as well as in declination, according to the results of the Astrolabe Catalogue. However, the data of this catalogue seem to indicate that FK4 system of declinations is more reliable than that of the right ascensions.

References

Anguita, C., Zverev, M. S., Carrasco, G., and Polozhentsev, D. D.: 1971, *Izv. Glav. Astron. Obs. Pulkovo*, No. 189–190, 83.
Anguita, C., Zverev, M. S., Carrasco, G., Loyola, P., Naumova, A. A., Nemiro, A. A., Polozhentsev, D. D., Taibo, R., and Shishkina, V. N.: 1973, *Inf. Bull. South. Hemisph.*, No. 22, 28.
Brosche, P. 1966, *Veröff Astron. Rechen-Inst. Heidelberg*, No. 23.
Loyola, P. and Shishkina, V. N.: 1968, *Izv. Glav. Astron. Obs. Pulkovo*, No. 183, 25.
Noël, F.: 1968, *ESO Bull.*, No. 4, 9.
Zverev, M. S.: 1969, *Astron. Zh.* **46**, 1290.

DISCUSSION

Fricke: I want to congratulate the team of observers from the Pulkovo and the Santiago Observatory on their fine results. Dr Anguita has mentioned that in view of the large deviations they found with respect to the FK4 it appears necessary to compile the SRS catalogue in an improved system. A decision on this question will be made in one or two years from now after consultation between the workers in Santiago-Pulkovo, Washington and Heidelberg. In my opinion it would certainly be best if the SRS catalogue could be given in the system of the FK5.

Wall: Any additional optical positions for quasi-stellar objects would be helpful in the position calibration of the Parkes telescope. Approximately how many such positions are you measuring?

Anguita: About twenty.

Luyten: I hope you will keep these plates taken with the Maksutov telescope on Quasars very carefully and that you will repeat them in 20–25 years. For then, if you measure the proper motion of the quasar relative to, say, 10–20 stars of about the same magnitude, you will get exactly what we have always wanted: the correction from relative to absolute proper motion for stars of a given magnitude.

Anguita: Yes, we will keep these plates on Quasars as carefully as the first epoch plates of the Proper Motion program observed with the Maksutov Astrograph, and we intend to repeat them in 20 years with the same purpose you mentioned.

Gliese: The mean epoch of the right ascension series observed at Cerro Calán is about 1966 or somewhat later. The small Cape polar catalogue observed in 1936 shows already nearly the same large $\Delta\alpha_\delta$ errors of the FK4 system which means that no significant proper motion corrections can be derived from that epoch difference of about 30 years and that, probably, errors in the fundamental proper motion system alone will not be responsible for that large $\Delta\alpha_\delta$ error in FK4.

Anguita: I agree with you. It seems to me that the large $\Delta\alpha_\delta$ errors in the southern part of FK4 in the 1960's are due mainly to the system of positions of the FK4 at the mean epoch of the catalogue and probably only a small fraction of them to the fundamental proper motion system. Probably this is what we may expect if the FK4 system in the southern part of the sky is based mainly on observations made with only one instrument: the Cape Transit Instrument.

PRELIMINARY RESULTS OF RELATIVE DECLINATION OBSERVATIONS MADE WITH THE REPSOLD MERIDIAN CIRCLE AT CERRO CALÁN

G. CARRASCO

Departamento de Astronomía, Universidad de Chile, Chile

Abstract. Preliminary results obtained from 6 years of declination observations are presented. Some instrumental details are discussed, and the observed declination system is compared with FK4 in $\Delta\delta_\delta$.

During the observations of the Southern Reference Stars and Bright Stars Programme with the Repsold Meridian Circle ($d = 19$ cm, $F = 224$ cm) at Cerro Calán Observatory, series of observations were made, along the whole meridian arc, of stars of the Fundamental Catalogue FK4. This paper gives the preliminary results in declination of the observations made from January 1963 to December 1968.

Four observers carried out 52 series, with a total number of 1721 individual observations of stars of the FK4 Catalogue with declination from $+41°$ to $-90°$ in upper culmination and from the pole to $-68°$ in lower culmination (Table I). This material represents about 40% of the total number of observations made by ten observers during this period.

About 20% of the series were made with visual circle readings and for the other 80%, FOLM photographic cameras (four on each side) were used to take photographs of the circle readings.

All the films were measured by the author in the Pulkovo Observatory, with the semi-automatic photo-electric measuring machine designed by Dr Platonov.

Before the start of the observations, the screw of the eye-piece micrometer and the screws of the four microscope-micrometers for the visual circle reading were investigated by the chain method. For the determination of periodic errors, the Rydberg method was used with an interval of $\frac{3}{5}$ and $\frac{3}{4}$ of revolution. For the eye-piece micro-

TABLE I

Distribution of the observations by observers

Observer	Number of series	Number of observations			Mean errors	
		UC	LC	Total	E_0	Observation
Anguita	8	198	29	227	$\pm 0\rlap{.}''476$	$\pm 0\rlap{.}''497$
Carrasco	31	895	151	1046	0.449	0.541
Loyola	8	283	29	312	0.452	0.542
Mercado	5	123	13	136	0.516	0.560
Totals	52	1499	222	1721	$\pm 0\rlap{.}''473$	$\pm 0\rlap{.}''535$

Gliese, Murray, and Tucker, 'New Problems in Astrometry', 73–78. All Rights Reserved.

meter screw the periodic error has a maximum value of $0''.07$ (equal to $0.004\ R$). The progressive error is practically the same for all the screws. The screws of the micro-scope-micrometers show a very small periodic error.

The value of one revolution of the declination screw was determined several times during these years with the mercury mirror. The value adopted was $R = 18''.38$.

For the investigation of the diameter errors of the main circle the Bruns method was used. The corrections of the circle reading were determined for an interval of $3°$ using three rosettes $R(3, x)$, $R(4, x)$ and $R(5, x)$. The systematic error is about $0''.4$ and the individual corrections go from $+0''.45$ to $-0''.35$ for the four microscopes. The centring error of the main circle corresponds to about $19''$ of the circle and its in-clination from the plane normal to the rotation axis is $0''.9$ (Anguita et al., 1963).

A second investigation of the corrections of all diameters (each $4'$), using the Bruns-Levy photographic method (Zverev, 1964) was carried out at the end of 1966 according to the following schema:

$$R(2700, x) = R(45, x) \times R(6, x) \times R(10, x).$$

For this purpose, four photographic cameras installed in the East cage were used, in which the normal indexes were replaced by new ones with 12 divisions at a spacing equal to that of the circle graduations.

The main object of the investigation, the measure of the rosette $R(45, x)$, was done with six angles between the diameters ($36°$, $72°$, $20°$, $60°$ and $80°$) and for the reduction of all the measures to the main system, two sets of exposures were taken; for the first coordination, angles $28°40'$, $29°20'$ and $30°00'$ were used, and for the second coordina-tion, $67°28'$, $67°36'$ and $67°40'$.

Nearby 4000 exposures (16000 pictures) were taken and the measures of this film, with the new automatic measuring machine designed in Odessa, are in progress in the Pulkovo Observatory.

The reductions of the observations were made with the IBM-360 computer of the University of Chile, using the following formulae:

$$E = \bar{M} + \overline{\Delta m}_{\mathrm{div}} - R''_\delta (\bar{m}_\delta - m_0) \pm \overline{\Delta m}_k \pm M_\varrho \pm \delta,$$

where the upper sign is for clamp East and the lower for clamp West; E, is the equator point; $\bar{M}$, the mean reading for the four microscopes; $\overline{\Delta m}_{\mathrm{div}}$, the correction for the circle reading (preliminary results obtained in 1962); R''_δ, the values of one revolution for the declination screw; $\bar{m}_\delta$, mean of the four readings of the declination screw; $m_0 = 4000$; $\overline{\Delta m}_k$, correction for the curvature of parallel; M_ϱ, correction for refrac-tion (computed from the Pulkovo Refraction Tables) and δ, the apparent declination of the FK4 stars. This formula was used for upper culmination; for lower culmination δ was replaced by $|\delta| - 180°$.

For each series the equator point E_0 was determined with stars whose zenith distance is $|Z| < 30°$. More than 80% of the observations show a large rate of the equator point with time. With the zenith stars a linear rate of the equator point was determined and with these parameters the equator point E_0 for each observed star is computed and the difference $\Delta E = E - E_0$ is obtained.

Only three series show anomalies in the rate of the equator point with time and they were reduced by a graphical method.

On the average, the mean error of one determination of the equator point E_0 is $\pm 0''462$ and the mean error of one observation is $\pm 0''540$.

The differences ΔE were grouped in zones of $5°$ and the mean values $\overline{\Delta E}$ for each zone were computed. All the observations with ΔE differing by more than $1''00$ from the mean value were excluded after a careful examination of the observing data. The mean error for one difference ΔE is $\pm 0''463$ for upper culmination and $\pm 0''500$ for lower culmination.

These values, slightly smoothed, are given in Table II for the two positions of the instrument, and are plotted in Figure 1. This figure represents 908 values for clamp E and 637 for clamp W.

TABLE II

Mean values $\overline{\Delta E}$ in both positions of the instrument, for $5°$ declination zone
(Unit: $0''01$)

δ	$\overline{\Delta E}$		δ	$\overline{\Delta E}$	
	E	W		E	W
$+40°$	$+18$	-42	$-40°$	-18	-10
$+35$	-07	-39	-45	-04	-11
$+30$	$+07$	-35	-50	-06	$+03$
$+25$	$+40$	-13	-55	-17	-08
$+20$	$+34$	$+17$	-60	$+04$	$+07$
$+15$	$+38$	$+15$	-65	-07	$+43$
$+10$	$+43$	-10	-70	-40	$+63$
$+05$	$+30$	-14	-75	-31	$+68$
00	$+16$	-08	-80	$+01$	$+50$
-05	$+08$	-02	-85	-05	$+18$
-10	$+16$	$+03$	-90	-32	$+30$
-15	$+07$	$+10$	-85 sp	-48	$+42$
-20	-04	$+04$	-80 sp	-31	$+35$
-25	$+01$	-03	-75 sp	-38	$+10$
-30	$+06$	-05	-70 sp	-12	-02
-35	-04	$+02$			

In this figure it is possible to see the rate in declination of the ΔE and also the horizontal flexure, given by $f = b \sin z$; there is a large deviation in the clamp W position between $-65°$ to $-80°$, probably due to errors in the first determination of the circle diameters corrections; this deviation appears for clamp E in the symmetrical position in zone $+5°$ to $+20°$.

In order to determine if this deviation is real for the Repsold Meridian Circle, the ΔE values were computed for the main observers: the results are given in Table III. The points in the zone $-65°$ to $-80°$ for clamp W show very good agreement.

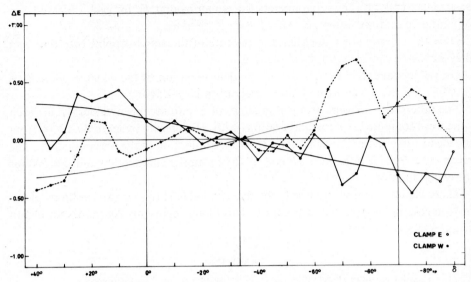

Fig. 1. Mean values $\overline{\Delta E}$ for the two positions of the instrument. (Slightly smoothed.)

TABLE III

Mean values $\overline{\Delta E}$ in both positions of the instrument by observers
(Unit: 0".01)

δ	Clamp E			Clamp W	
	Anguita	Carrasco	Loyola	Carrasco	Loyola
+40°	−08	+32	−	−51	−
+35	−29	+01	−39	−50	−32
+30	+29	+19	−37	−52	−43
+25	+32	+54	−05	−37	00
+20	+22	+38	+20	−11	+37
+15	+39	+34	+43	−02	+29
+10	+45	+41	+24	−13	00
+05	+30	+28	+18	−26	+15
00	+07	+11	+35	−24	+27
−05	+14	+02	+25	−17	+07
−10	+45	+07	+16	−02	+01
−15	+33	+06	−10	+14	+29
−20	−04	+01	−18	+04	+20
−25	00	+03	+06	−01	−09
−30	+17	+08	−06	−01	−09
−35	+03	−02	−28	+04	−07
−40	−19	−17	−17	−07	−10
−45	−11	−04	+03	−13	+02
−50	−12	−11	+25	+02	+16
−55	−46	−14	+05	−06	−03
−60	−35	+12	+22	+10	+06
−65	−10	−14	+22	+48	+35
−70	−21	−49	−28	+67	+64

Table III (Continued)

−75	−36	−28	−35	+69	+70
−80	−13	+04	−02	+50	+48
−85	−09	−02	−14	+02	+25
−90	−33	−38	+15	+29	−03
−85 sp	−30	−53	−	+53	+33
−80 sp	−48	−30	−	+45	+42
−75 sp	−56	−40	−	+13	+11
−70 sp	−	−14	−	−17	+25

TABLE IV

Systematic differences of the type $\Delta\delta_\delta$ in the
sense Repsold Meridian Circle – FK4
(Unit : 0″01)

δ	$\Delta\delta_\delta$	δ	$\Delta\delta_\delta$
+40°	−12	−40°	−14
+35°	−23	−45°	−08
+30°	−14	−50°	−02
+25°	+14	−55°	−13
+20°	+25	−60°	+05
+15°	+26	−65°	+18
+10°	+16	−70°	+11
+05°	+08	−75°	+19
00	+04	−80°	+26
−05	+03	−85	−06
−10	+10	−90	−01
−15	+08	−85 sp	−03
−20	−02	−80 sp	+02
−25	−01	−75 sp	−14
−30	+01	−70 sp	−07
−35	−01		

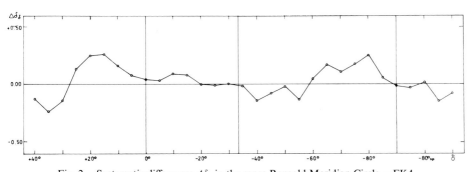

Fig. 2.   Systematic differences $\Delta\delta_\delta$ in the sense Repsold Meridian Circle −FK4.

Finally the $\frac{1}{2}(E+W)$ values were computed, representing the systematic differences $\Delta\delta_\delta$ in the sense Repsold Meridian Circle − FK4. The results are given in Table IV and Figure 2. The deviation of the curve in the southern declinations is probably real. For northern declinations this deviation is probably due to refraction anomalies.

Acknowledgement

The author wishes to thank the staff of the Astronomy Department, Observatorio Astronomico, Cerro Calán, for their help in this work.

References

Anguita, C., Bagildinsky, B. K., Carrasco, G., Loyola, P., Shishkina, V. N., and Zverev, M. S.: 1963, *Inf. Bull. South. Hemisph.* **3**, 30.
Zverev, M. S.: 1964, *Astron. Zh.* **41**, 1128 (1965, *Soviet Astron.* **8**, 897).

DISCUSSION

Wood: The results of Dr Carrasco reveal an important point about the surroundings in which a transit instrument should be sited.

Dieckvoss: Did you measure the circle photographically all the time?

Carrasco: The circle was measured visually for 20% of the observations, and photographically for the other 80%.

The general results in both cases are equal.

OBSERVATIONS BY THE HAMBURG SRS-EXPEDITION
TO PERTH

ERIK HØG

Hamburger Sternwarte, F.R. Germany

and

I. NIKOLOFF

Perth Observatory, Western Australia

Abstract. A total of 110000 observations of 26000 stars was obtained between 1967 and 1972. The internal mean errors are $0''.17$ and $0''.27$ in $\Delta\alpha\cos\delta$ and $\Delta\delta$, respectively, both multiplied by the factor $(\cos z)^{-0.5}$ at the zenith distance z. The preliminary evaluation also shows that the internal systematic errors, judged from the clamp corrections in $5°$ zones in declination, are less than $\pm 0''.02$ and $\pm 0''.15$, respectively. The systematic differences in right ascension from the FK4 are given.

1. Introduction and Conclusions

The Hamburg expedition conducted by J. von der Heide enjoyed the hospitality of, and a close collaboration with the Perth Observatory for many years. The observations of the SRS program were started in November 1967. After the departure of the eight German observers in December 1971 I. Nikoloff continued the observing program until August 1972 when about 110000 observations of R.A. and Decl. had been obtained during 580 nights.

The observations encompass 24803 observations of FK4 stars with $\delta < +38°$, at least 4 observations of all stars with $\delta < -40°$ and at least 3 observations for 80% of all program stars in the SRS catalogue. About 8% of the catalogue stars have only one observation and a few hundred have not been observed at all. Every observation of a star gave a magnitude m_v with a mean error of $0''.09$. In addition 560 observations of major and minor planets were obtained.

The photoelectric Repsold Meridian Circle, which is still in operation at Perth on a program of FK4 Supp stars, has been described by Høg (1972) and a preliminary analysis of 150 nights of observations has been given.

The following contains a more detailed analysis of the observations of the FK4 stars in right ascension and finally the difference (Cat. $-$ FK4) for $\Delta\alpha_\delta\cos\delta$. Further checks on the internal and external accuracy are still being carried out, and in this sense the present analysis is only preliminary. As far as yet the following conclusions are justified:

(1) The clamp corrections for $5°$ zones are about $\pm 0''.010$.

(2) The internal mean error of one observation of a star is $0''.18$ and the asymptotic mean error is less than $0''.030$.

(3) The values of $\Delta\alpha_\delta\cos\delta$ in Figure 3 for $5°$ zones have internal systematic errors less than $\pm 0''.020$ and perhaps equal to only $\pm 0''.010$.

Gliese, Murray, and Tucker, 'New Problems in Astrometry', 79–84. All Rights Reserved.

(4) A photoelectric multislit micrometer improves not only the accidental errors but also the systematic errors, probably by diminishing the thermal and other influences from the observer on the whole instrument.

Thus, the performance of the photoelectric Repsold meridian circle is even better than the first 150 nights indicated. The high systematic accuracy has not been known before, and it should encourage the modernization of conventional meridian circles with new micrometers, circles and pivots, since a lot can be gained without introducing new types of meridian circles (Høg, 1973).

The accuracy of the observations in Decl. is better than of visual observations, the mean error being 0″.27 and the systematic errors less than 0″.15. The lesser accuracy in Decl. than in R.A. is caused by the imperfect circles and by our incomplete knowledge of the division errors.

2. Right Ascensions of FK4 Stars

The average observing night lasted 9 h and contained 4 observations of nadir and marks, 160 program stars and 43 FK4 stars. The FK4 stars were distributed equally during the night over an interval of 7^h of R.A. and over the zenith distances $\pm 70°$. The observations in R.A. of the night were corrected for the errors of collimation and drifts in inclination and azimuth by means of the observations of nadir and marks. The FK4 stars therefore served only to determine two constant parameters for a night, i.e. the rotation around two axes: azimuth and clock correction. The mean errors for the two parameters were 0″.10 and 0″.06, respectively, for a night.

The observations were designed to be relative to the FK4, but since they are linked so firmly to a large area of the sky every night, they can reveal certain systematic errors of the FK4. Attempts to observe the same stars in upper and lower culmination in one night were not successful due to the low latitude; neither were the nights linked together by assuming a constant direction to the marks from night to night as is usual in absolute observations.

A systematic correction of unknown origin to the collimation of about 0″.2, and constant over a few years, was introduced in order to remove part of the clamp differences. The resulting clamp corrections, i.e. half of the difference, is shown in Figure 1c, as derived from 24 803 observations of FK4 stars. The pivot corrections derived from direct measurement on the pivots and shown in Figure 1a were *not* applied. Had they been real they would have caused the clamp corrections shown as '$\frac{1}{2}$(E − W) pivots' in Figure 1b. Obviously, however, the observed clamp corrections are smaller and show no similarity to the plot above. This proves that the pivots are extremely good with irregularities generally less than 0.05 μm. Actually the pivots are elliptical with an ellipticity about 0.15 μm for both pivots. The axes of these ellipses are nearly parallel; whether by accident or by virtue of the manufacturer (Heidenreich und Harbeck, 1964) we do not know. Therefore, an ellipticity effect of the form $\sin(2z + \psi)$ does not appear in R.A.

For comparison the pivot and clamp corrections for the Washington six-inch

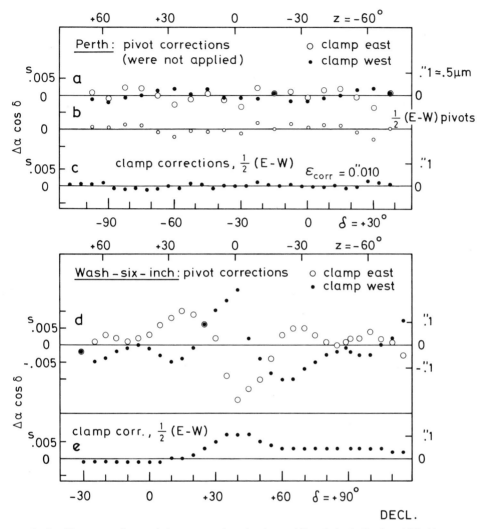

Fig. 1. Pivot corrections and clamp corrections for the meridian circles in Perth and Washington.

(Adams and Scott, 1968) are shown in Figures 1d and 1e. The rapid variation of the pivot corrections near zenith has apparently not been measured quite correctly so that larger clamp corrections remain in that region. The values in Figure 1e are smoothed between the 5° zones, and their scatter therefore cannot be compared with the unsmoothed values in Figure 1c.

The residual scatter in Figure 1c is most simply expressed as $\varepsilon_{corr} = 0\rlap{.}''010 = 0\rlap{.}^s0007$ for a 5° zone. Since each point is derived from about 400 observations in each clamp position we, therefore, have $\varepsilon_{400} = 0\rlap{.}''014$, where ε_n is the mean error for the mean value of n observations. This error must be due to pivot irregularities, image motion, etc.; the mean error $\varepsilon_1 = 0\rlap{.}''18$ for one observation contributing $0\rlap{.}''18/400^{0.5} = 0\rlap{.}''009$.

The internal mean errors for one observation are shown in Figure 2. The mean errors in R.A. from the 24 603 observations within the $\pm 3\sigma$ limits do not differ significantly whether observations from both clamp positions are taken together or whether they are treated separately, see Figures 2b and 2a – in agreement with the minute clamp corrections (which are not applied). The mean error increases less with the zenith distance than predicted by the theoretical law $(\cos z)^{-0.5}$ and may be represented by the value $\varepsilon_1 = 0\overset{''}{.}18 = 0\overset{s}{.}012$ at all zenith distances $z \leqslant 50°$.

Figures 2c and 2d show the mean errors for Decl. which have a dependence on zenith distance nearly as predicted by the theoretical law. The values in Figure 2d where both circles have been used for each star are much influenced by our incomplete knowledge of the division errors while Figure 2c is less dependent on this effect.

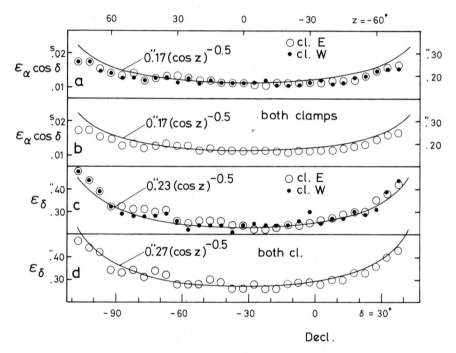

Fig. 2.   Internal mean errors for one observation as function of zenith distance based on 24 803 observations of FK4 stars. The mean errors are given for R.A. and for Decl. and for each of these separated in clamp east, clamp west and both clamp positions together.

The statistical error distribution* shall be given since it is a measure for the reliability of a catalogue. In this connection we mention the study by Dejaiffe (1973). The distribution of deviations of the single observations of R.A. from the mean values were found to differ significantly from a Gaussian in the wings (but probably less so

---

* The present values for the error distribution differ somewhat from the values given in the preprint distributed at the IAU Symposium. Also the mean errors at large zenith distances are slightly changed – both due to the correction of two errors in the computer program.

than other meridian observations). The percentages of deviations outside the limits $\pm n\sigma$, where $\sigma = 0''.18(\cos z)^{-0.5}$ and $n = 1, 2, 3, 4, 6$ and 9, were 27.9 (31.8), 4.5 (4.6), 0.81 (0.27), 0.28 (0.006), 0.13 (0.000), and 0.06 (0.000), respectively. Values for a Gaussian are given in brackets. The percentages were derived from 24803 FK4 observations, and the 0.81% of the observations outside $\pm 3\sigma$ were rejected. Out of 85088 observations of program stars 0.74% deviations are outside $\pm 3\sigma$. It is statistically important for small numbers of observations n of a star to decrease the width of the limits by the factor $(1 - (1/n))^{-0.5}$, and this has be done throughout.

Returning to ε_n for R.A., other estimates were obtained from differences between observations during the first 2.5 years and the last 2.5 years giving $\varepsilon_{200} \leqslant 0''.038$. Finally, clamp differences for individual stars gave $\varepsilon_{10} = 0''.064$. These three values of ε_n lead to asymptotic mean errors $0''.011 \leqslant a \leqslant 0''.035$ when the formula

$$\varepsilon_n^2 = a^2 + (b^2/n) \tag{1}$$

is applied.

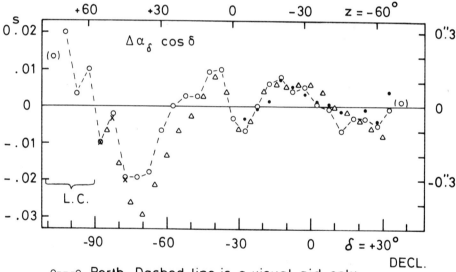

Fig. 3. Differences in right ascension from observations in Perth, Washington and Santiago in the sense (Cat − FK4).

This may be compared with the asymptotic mean error $a = 0''.084$ for R.A. in modern visual catalogues given by Kopff et al. (1964).

Figure 3 shows the values of $\Delta\alpha_\delta \cos\delta$ for 5° zones without any smoothing between zones. Values are given for comparison from Washington (Adams and Scott, 1968) and Santiago/Pulkovo (Zverev, 1970). The values from Santiago are based on 7420 visual observations and have been somewhat smoothed between zones by their authors. Good agreement between the three sources exists in confirming the systematic errors of FK4 notably around $\delta = -70°$, but also in the whole interval up to $\delta = +30°$. It is doubtful whether a new reduction of all nights with corrected positions of the FK4 stars would change our results significantly, since so many well distributed FK4 stars were used every night – on the average 43 – and in different combinations with program stars. The systematic difference of $0''.10$ between the observations from Perth and from Santiago in the interval $-75° \leqslant \delta \leqslant -50°$ is statistically significant, but of quite common magnitude for visual observations.

The accuracy of the Perth observations is confirmed by the observations in lower culmination which differ from the values obtained for the same stars in upper culmination by $(UC - LC) = 0''.00$, $+0''.02$ and $+0''.01$, respectively, for the zones centered at $\delta = -87°.5$, $-82°.5$ and $-77°.5$. The zone observed at $z = 75°$, see Figure 3, does not contain all stars in FK4 and can, therefore, not be used for comparison although the accuracy at that zenith distance is still acceptable.

References

Adams, A. N. and Scott, D. K.: 1968, *Publ. U.S. Naval Obs., Ser. 2* **XIX**, Part II.
Dejaiffe, R. J.: 1973, *Astron. Astrophys.* **22**, 425.
Høg, E.: 1972, *Astron. Astrophys.* **19**, 27.
Høg, E.: 1973, this volume, p. 243.
Kopff, A., Nowacki, H., Strobel, W.: 1964, *Veröff. Astron. Rechen-Inst. Heidelberg*, Nr. **14**.
Zverev, M. S.: 1970, *Izv. Glav. Astron. Obs. Pulkovo*, No. 185, 3.

DISCUSSION

Fricke: I may draw the attention of the audience to the fact that the results obtained in Perth are in marvellous agreement with those obtained by the Santiago-Pulkovo team at Santiago de Chile. This strengthens very much the corrections to be applied to the FK4 system. I wish our colleagues at Perth every success for the continuation of their work.

ON PROGRESS AND METHODS OF REDUCTION OF THE MELBOURNE OBSERVATIONS OF FUNDAMENTAL STARS

K. N. TAVASTSHERNA

Pulkovo Observatory, U.S.S.R.

The astronomers C. Merfield, J. Moroney, G. Wootthouse, J. Feely and W. Holmes of the Melbourne Observatory made a large series of observations of fundamental stars with the 8″ meridian circle (Troughton and Simms) during 1928–1941. After closing the Observatory all the materials concerned with this large work (about 20000 observations in δ and 30000 observations in α) were forwarded to the Mount Stromlo and then the Pulkovo Observatories for the final reduction and publication (Tavastsherna, 1963).

The reduction of declinations has now been completed. Four absolute catalogues have been published (Tavastsherna and Shakht, 1968). These catalogues include: (1) 2249 Backlund-Hough list stars from $+32°$ to $-90°$ of declinations ($Me_{50}I\delta$), (2) 144 latitude stars in the Kimura list ($Me_{50}II\delta$), (3) 35 stars with the $+32°--90°$ declinations from the A. Kopff list of 'Ersatzsternen' ($Me_{50}III\delta$) and (4) 38 zenith stars of the Melbourne Observatory ($Me_{50}IV\delta$).

At present the first stage of the reduction of right ascensions of these stars has also been completed. The differential catalogue was compiled in the N 30 system. We propose to compile an absolute catalogue using a combination of the step-by-step, chain and Greenwich methods.

References

Tavastsherna, K. N.: 1963, *Trudy 15th Astrometr. Konf.*, p. 156.
Tavastsherna, K. N. and Shakht, N. A.: 1968, *Catalogues of Stellar Positions Compiled at Pulkovo Observatory on the Basis of Observations made at Melbourne Observatory from 1928 to 1941*, Pulkovo, Leningrad.

Gliese, Murray, and Tucker, 'New Problems in Astrometry', 85. All Rights Reserved.

CONTRIBUTION DES ASTROLABES A LA DETERMINATION DU SYSTEME DE REFERENCE DANS L'HEMISPHERE SUD

G. BILLAUD and S. DEBARBAT

Observatoire de Paris, France

Abstract. The potentialities of astrolabe observations which are available for improving the fundamental system, particularly in the southern hemisphere, are discussed. Difficulties arise because of the poor distribution in latitude of the observing stations and diversity of observational programmes.

1. Généralités

L'observation répétée des mêmes étoiles met en évidence des erreurs systématiques dues, pour une large part, aux erreurs du catalogue fondamental et leur correction est l'un des buts que poursuivent les astronomes des deux hémisphères. Ils disposent pour cela de divers instruments; parmi ceux-ci, bien qu'il ne puisse atteindre les étoiles au-delà de la magnitude 6,0, l'astrolabe permet l'étude de 91% des étoiles du FK4 et de 55% de celles du FK4 Supp, avec une précision (écart-type d'une mesure en distance zénithale) de l'ordre de 0″.17 en moyenne. A ce titre, il est susceptible d'apporter une contribution importante à l'amélioration du catalogue fondamental.

Les conditions d'élaboration d'un catalogue astrolabe sont bien connues (Débarbat et Guinot, 1970). De tels catalogues sont en principe dépourvus d'équations de couleur et de magnitude, ainsi que d'erreurs du type $\Delta\alpha_\alpha$, $\Delta\alpha_\delta$, $\Delta\delta_\alpha$. Il subsiste toutefois:

(a) Une erreur constante des ascensions droites.

(b) Une erreur systématique des déclinaisons du type $\Delta\delta_\delta$ de forme connue, d'amplitude faible mais indéterminée.

Dans ces catalogues la précision en ascension droite est proportionnelle à $1/\cos\varphi \sin a$ et celle en déclinaison à $1/\cos S$ (φ, a, S désignent respectivement latitude, azimut et angle parallactique). Ceci entraîne que pour une latitude donnée, il existe, en ascension droite, deux zones de même précision, l'une vers le pôle, l'autre vers l'équateur, condition que l'on ne trouve pas dans les catalogues établis avec un instrument méridien où la précision dépend de $\cos\delta$. La même chose existe en déclinaison, où il faut remarquer la présence d'une zone d'indétermination qui couvre de 10 à 15 degrés en δ.

2. Résultats obtenus

Les catalogues réalisés ont montré que la méthode conduisait à des résultats sûrs et les craintes d'introduction d'erreurs supplémentaires dues à des anomalies de réfraction se sont montrées non fondées. Cependant, les résultats actuels ne justifient pas l'espoir que l'on avait eu lors de la première publication sur ce sujet (Guinot, 1958), malgré la très forte concentration d'astrolabes dans l'hémisphère nord, les résultats

Gliese, Murray, and Tucker, 'New Problems in Astrometry', 87–90. All Rights Reserved.

sont loin de correspondre aux possibilités réelles. Selon toute vraisemblance, cela tient au fait que l'astrolabe devait fournir en premier lieu heure et latitude et que les corrections aux positions du FK4 n'apparaissaient que comme un sous produit du programme fondamental, lui-même inadapté à l'établissement d'un catalogue: lorsqu'un programme spécial est absent, le nombre d'étoiles observées est insuffisant pour une étude systématique; de plus, cette insuffisance est fréquemment aggravée par le manque d'homogénéité des programmes d'observation. Les zones de recouvrement existent, mais il est exceptionnel qu'elles contiennent plus de 50 étoiles, ce qui est très peu.

Le tableau ci-dessous donne les valeurs des coefficients de recouvrement, c'est-à-dire le nombre d'étoiles communes à deux catalogues par degré de déclinaison; les valeurs inscrites sur l'hypothénuse du triangle fournissent la densité du catalogue.

Les observatoires mentionnés couvrent de $-60°$ à $+77°$ en déclinaison.

TABLEAU I

	Pt	H	P	N	W	A	C	Q	T	S.J.
Potsdam	2,1									
Herstmonceux	0,9	2,5								
Paris	1,8	2,1	6,0							
Neuchatel	0,7	0,8	1,6	2,3						
Washington	0,4	0,5	1,1	0,4	1,5					
Alger	0,2	0,3	1,0	0,1	0,3	1,7				
Curaçao	0,4	0,4	1,0	0,2	0,0	0,2	1,3			
Quito	1,0	1,9	2,5	0,9	0,4	0,2	0,3	2,9		
Tananarive							0,7	0,2	0,6	1,9	
San Juan								0,6	0,4	0,1	1,8

Dans l'hémisphère sud on constate les mêmes difficultés que dans l'hémisphère nord, accrues par une couverture encore plus faible. A cet égard on ne peut que regretter l'interruption des observations au Cap $(-33°9)$ et à Wellington $(-41°3)$.

3. Projets pour l'hémisphère sud

L'exploitation des résultats des deux stations temporaires de Curaçao $(+12°)$ et de Tananarive $(-19°)$ (Année Géophysique Internationale) a permis la publication de deux petits catalogues de 77 et 113 étoiles respectivement. L'observatoire de Quito $(0°)$ a, de son côté, publié un catalogue de 202 étoiles: il est inutile d'insister sur son importance dans le raccordement des deux hémisphères.

A ces publications, devraient se joindre bientôt le catalogue de Santiago du Chili $(-33°)$, de San Juan $(-32°)$, le second catalogue de l'observatoire de Quito ainsi que le catalogue de l'observatoire du Cap et peut-être celui de Wellington.

Ces perspectives encourageantes montrent tout l'intérêt que les astronomes de l'hémisphère sud portent à ces travaux. Il convient de noter à ce sujet une entreprise originale des astronomes d'Amérique Latine qui ont suscité en novembre 1971, à l'occasion du centenaire de l'observatoire de Cordoba (République Argentine), le 'First Latin American Astrometric Working Meeting'. Au cours de cette réunion, où l'on s'est tout particulièrement penché sur les problèmes d'homogénéisation et de coordination, plusieurs recommandations ont été approuvées.

Outre les prochaines publications de catalogues que nous avons déjà signalées s'inscrivent, dans les projets à court terme, la mise sur pied de stations dans l'extrême sud américain ($-53°$ et $-46°$). Ces projets doivent être fortement encouragés, car si les conditions climatiques le permettent, l'élaboration d'un catalogue, couvrant en déclinaison une zone allant de $-20°$ à $-80°$, apporterait à la connaissance du ciel austral une contribution essentielle.

L'importance des erreurs du catalogue fondamental pour les déclinaisons négatives justifie la publication de catalogues provisoires établis à partir d'un moins grand nombre d'observations que dans l'hémisphère nord ; le problème est en effet analogue à celui qu'a traité le Professeur Lacroute lors du Colloque no. 20 à Copenhague dans un domaine voisin. Concrètement il apparait qu'avec un an d'observations de qualité moyenne, correspondant à un écart quadratique moyen de $0\rlap{.}''25$, on disposerait d'un catalogue préliminaire intéressant, mettant en évidence les erreurs du FK4 d'une manière significative.

4. Projet d'un catalogue général astrolabe

Il reste ensuite à raccorder entre eux les différents catalogues des deux hémisphères. Les études faites à Paris ont montré la possibilité, à partir des corrections des étoiles communes, de réaliser un catalogue homogène. Ce catalogue est en cours d'exécution ; il est pratiquement achevé en ce qui concerne les α. Il comprendra en particulier un fichier où seront répertoriées les contributions des divers observatoires, leurs écarts systématiques ainsi que le jeu des corrections d'harmonisation de chacun d'eux. La révision lors de l'adjonction d'un nouveau catalogue ne présentera aucune difficulté.

La qualité du raccordement sera grandement améliorée si une nouvelle station équatoriale peut être créée. Cette création aurait des conséquences importantes non seulement pour les catalogues mais aussi pour la détermination du temps et de la latitude.

D'autres possibilités sont ouvertes. Krejnin (1968) a en effet montré qu'il était possible d'obtenir, en associant deux astrolabes travaillant à des distances zénithales différentes, un catalogue de déclinaisons absolues. Un tel programme de recherche pourrait être poursuivi dans les prochaines années en utilisant l'astrolabe à pleines pupilles (Billaud et Guinot, 1971), dont il existe à l'Observatoire de Paris un prototype en service permanent. Deux autres instruments actuellement en cours d'étude seront aussi adaptés à l'emploi de la méthode de Krejnin : il s'agit du Reflecting Astrolabe (Herstmonceux) et de l'Astrolabe Photoélectrique (CERGA, Grasse).

Comme on l'a vu, les observations d'étoiles à l'astrolabe ne permettent pas de fixer l'équinoxe : cela ne peut être fait directement qu'en observant le Soleil et une telle entreprise présente de grandes difficultés. Il existe cependant une méthode indirecte de rattachement: c'est l'observation des planètes. Aux observatoires de l'hémisphère nord: Alger, Besançon, Paris, San-Fernando se sont déjà joints les observatoires de Quito et de Sao Paulo, auxquels se joindront bientôt ceux de San Juan et de Santiago du Chili, permettant l'observation de Mars, Jupiter, Saturne, Uranus et Vesta ainsi qu'il a été recommandé par le colloque UAI no. 20 (Copenhague 1972). Une collaboration internationale est nécessaire pour assurer la couverture des trajectoires avec le maximum de précision, celle-ci dépendant, selon la déclinaison, de la latitude du lieu d'observation.

5. Conclusion

Ainsi l'astrolabe apparaît comme un outil efficace pour l'amélioration du système de référence, tout particulièrement dans l'hémisphère sud. La tâche entreprise doit être poursuivie, et menée avec le souci de satisfaire aux nécessités des raccordements entre catalogues.

Bibliographie

Billaud, G. et Guinot, B.: 1971, *Astron. Astrophys.* **11**, 241.
Débarbat, S. et Guinot, B.: 1970, *La méthode des hauteurs égales en astronomie*, Gordon and Breach.
Guinot, B.: 1958, *Bull. Astron.* **22**, 1.
Krejnin, E. I.: 1968, *Astron. Zh.* **45**, 447.

DISCUSSION

Fricke: There is no doubt about the importance of the contributions obtained with astrolabes for the improvement of the fundamental system. Miss Débarbat's request for a more rational world-wide organisation of such observations is justified. I am sure that IAU Commission 8 will be willing to adopt a resolution to this effect.

SOME URGENT PROGRAMMES OF MERIDIAN OBSERVATIONS ESPECIALLY FOR THE SOUTHERN HEMISPHERE

(Invited Paper)

M. S. ZVEREV

Pulkovo Observatory, Leningrad, U.S.S.R.

Abstract. Recommendations of Commission 8 at previous IAU General Assemblies on the following programmes are recalled: Bright Stars (BS), Double Stars unsuitable for photographic observations (DS), Latitude Stars (LS) and Zodiacal Stars (ZS). One more programme – 'Additional Reference Stars in the areas with Galaxies of the Pulkovo Programme' (PS) containing 735 stars for the whole sky is proposed.

For the observers with meridian instruments there exist the following main problems:

(1) Observations of fundamental stars – bright (FK4) and faint (including FKSZ) – for improving the fundamental system of coordinates and proper motions. Absolute observations of these stars are of special importance.

(2) Regular observations of the Sun and planets (major and some of the minor ones) and also the Moon – for improvement of orbits and orientation elements for the fundamental system. Differential observations in the fundamental catalogue system are preferable.

(3) Differential determinations of coordinates of the reference stars for photographic astrometry. We think that the totality of the AGK3R + SRS stars can be named 'international reference stars' (IRS). This program contains about 41 000 stars for the whole sky, i.e. on the average one star per square degree.

Besides the above problems there are some other urgent tasks in Meridian Astronomy. Some of them were discussed at previous IAU Assemblies and the recommendations for observatories to participate in observations were adopted. However these recommendations have been insufficiently realized. I should like to recall some of them and to add one more program.

1. Bright Stars (BS)

Resolution No. 17 of the Xth Assembly of the IAU (Moscow, 1958) recommended making meridian observations of all bright stars up to 6^m00; the XIth IAU Assembly (Berkeley, 1961) recommended (resolution No. 7 of Commission 8) observing the southern bright stars together with the SRS. The working list of BS was composed at the Pulkovo Observatory and improved by Dr F. P. Scott (the Naval Observatory in Washington). For the whole sky it consists of 5115 stars. Several observatories (Strasbourg, Washington, Kiev, Tashkent, Kharkov, Nikolayev, Santiago, Tokyo and perhaps some others) undertook the meridian observations of these stars. In the U.S.S.R. these observations were made more intensively at the Kiev University Observatory under the guidance of Dr N. A. Chernega. He intends to compile the

Gliese, Murray, and Tucker, 'New Problems in Astrometry', 91–94. All Rights Reserved.

General Catalogue of BS using all the observations. In reply to our request we have already received the consent from several observatories; the Strasbourg, Tashkent and Nikolayev Observatories have sent the result of BS observations in the form of a final catalogue.

It seems expedient to approve the suggestion of the Kiev Observatory to compile the BS General Catalogue and to ask observatories to send the observational results to Kiev. In the future it is important to continue the meridian observations of all the BS stars, especially in the southern hemisphere, as was recommended in Resolutions of the above-mentioned General Assemblies of the IAU.

2. Double Stars (DS)

The resolution No. 17b adopted at the Xth Assembly recommends the meridian observations of the double stars unsuitable for photographic observations. The preliminary working list was compiled at the Pulkovo Observatory; it was improved and supplemented by Scott (1960). The list contains 2292 double stars for the whole sky. For each DS only the brighter star or the preceding one in R.A. is to be observed.

The observations according to this program were made apparently only in Strasbourg (the zones from +25° to +71° in declination) and in Santiago (the zones from −47° to −90°). Prof. P. Lacroute has kindly sent to Pulkovo the results of these observations.

Most of the stars from the DS program were not included in the AGK3 and other photographic catalogues. Study of their proper motions is important for Stellar Astronomy, but in the last 50 years their coordinates have apparently not been determined anywhere, except at the two above mentioned observatories. Thus the above recommendation remains entirely valid.

3. Latitude Stars (LS)

This program aims at the reduction of latitude observations, made at various observatories, to a uniform system, which would permit studying in detail a periodic and, later on, a secular polar motion and also non-polar latitude variations of observatories. The LS program has been compiled at the Sternberg Astronomical Institute (Fedorov et al., 1960) for the northern hemisphere. It contains about 3900 stars, 1300 of which are included in the programs of 11 PZTs, and require observations in both R.A. and declination. The remaining 2600 stars are included in the zenith telescope programs of about 20 observatories, and for these stars only declinations need be observed. The importance of meridian observations of the latitude stars is underlined in Resolutions of the X, XI, XII, XIII and XIV IAU Assemblies.

At present the declinations of LS have been determined at the Golosseyevo and Beograd observatories and are being determined at Pulkovo and in Moscow. At the latter two observatories the meridian determination of R.A. of about 1200 PZT stars have been also made. These observations are, of course, not enough for compiling the General Catalogue, therefore the need for participation of other observatories remains still valid.

I welcome the initiative of Japanese astronomers in organizing meridian observations of about 1700 PZT stars according to the list of Yasuda (1971), and wish success to this undertaking.

4. Zodiacal Stars (ZS)

The resolution (item 4) of Commission 8 of the XIV Assembly at Brighton (1970) recommends meridian observations of the Zodiacal stars according to the program compiled at the Nikolajev Observatory. This program contains 9518 stars, mainly brighter than $8^m.0$, in the zone of $\pm 15°$ from the Ecliptic.

The last General Catalogue of about 3500 zodiacal stars was published in Washington in 1940 (Robertson, 1940). Since then only a few meridian observations of these stars have apparently been made. In this connection the suggestion of the Nikolajev Observatory to make a new series of meridian observations of ZS seems to be quite timely, especially in view of modern cosmic researches.

5. Reference Stars in the Areas with Galaxies of the Pulkovo Program (PS)

The Pulkovo program of photographing the areas with galaxies (the Deutsch program) was suggested at the U.S.S.R. Astrometrical conference in November 1938 (Zverev, 1939, 1941). The working list of 271 areas for both hemispheres was published by Neuymin (1940). After looking through the numerous test photographs this list for the northern sky (as far as $-5°$ declination) was revised and extended at the Pulkovo Observatory (Deutsch *et al.*, 1955). The analogous work for the southern hemisphere was made at the Tashkent Observatory as far as $-25°$ declination (Latypov and Fatchikhin, 1955) and more southern declinations at the National Observatory of Chile (Gutierrez Alonso, 1961). Systematic observations according to this program have been made at the U.S.S.R. observatories since 1939. At present photographs of the first epochs have been taken at 12 observatories in both hemispheres with the long-focus and wide-angle astrographs.

The number of reference AGK3R and SRS stars (one star per square degree on the average) is not sufficient for photographs with long-focus instruments (field $\leqslant 2° \times 2°$). Therefore a list of additional reference stars in these areas for the northern sky was compiled at the Sternberg Institute (Moscow) in 1955 (Bugoslavskaya *et al.*, 1955). This list was extended to $-25°$ declination at the Tashkent Observatory (Fatchikhin, 1963) and recently extended to the south pole at the Pulkovo Observatory (Baturina, 1974).

The whole list of additional reference stars in 301 areas of the Pulkovo program contains 735 stars. Altogether 1630 reference stars (including the AGK3R and SRS) are in these areas, i.e. 5–6 stars per square $2° \times 2°$ on the average. These stars are to be observed with meridian instruments at the observatories in the northern and southern hemispheres.

It is desirable that the 735 additional reference stars should be included into the IRS = AGK3R + SRS program.

References

Baturina, G. D.: 1974, *Trudy 19th Astrometr. Konf.*, in press.
Bugoslavskaya, E. J., Karimova, D. K., Podobed, V. V., and Jakhontov, K. N.: 1955, *Trudy 11th Astrometr. Konf.*, p. 42.
Deutsch, A. N., Lavdovsky, V. V., and Fatchikhin, N. V.: 1955, *Izv. Glav. Astron. Obs. Pulkovo*, No. 154, 14.
Fatchikhin, N. V.: 1963, *Trudy 15th Astrometr. Konf.*, p. 161.
Fedorov, E. P., Prodan, Y. J., and Ponomarev, D. N.: 1960, *Trudy 14th Astrometr. Konf.*, p. 139.
Gutierrez Alonso, A.: 1961, *Publ. Obs. Nacional Chile*, No. 4.
Latypov, A. A. and Fatchikhin, N. V.: 1955, *Circ. Tashkent*, No. 302.
Neuymin, G. N.: 1940, *Kazan University Uchenye Zapiski* **100**, 86.
Robertson, J.: 1940, *Astron. Papers A.E.* **X**, Part 2.
Scott, F. P.: 1967, *Trans. IAU* **XIII A**, 73.
Yasuda, H.: 1971, *Ann. Tokyo Obs.* **8**.
Zverev, M. S.: 1939, *Astron. Zh.* **16**, 104.
Zverev, M. S.: 1941, *Astron. Zh.* **17**, 54.

REPORT ON THE CAPE PARTICIPATION IN THE SOUTHERN
REFERENCE STAR PROJECT

R. H. TUCKER

Royal Greenwich Observatory, Herstmonceux, England

The Gill Transit Circle at the Cape Observatory has observed zones $-40° -52°$ and $-30° -40°$ of SRS and is now observing zone $-52° -64°$, which is 80% complete. In addition to fundamental stars selected to define the FK4 system in the zone of observation, clock stars and azimuth stars are also observed so that the observations can be subsequently referred to a more independent or fundamental system.

Observations of the Bright Star list associated with SRS have also been made contemporaneously throughout the period (1961–1973).

The current reductions are being performed at Herstmonceux. Data forms completed and checked by hand at the Cape are sent to Herstmonceux, where they are encoded on magnetic tape by means of a Matador keyboard machine. Errors in data are corrected after consultation by correspondence, and the ledgered results are analysed and discussed in the Cape Meridian Department.

Preliminary investigation of the observations of FK4 stars for 1966–1969 shows that the instrumental declination system is practically free from clamp differences in $\Delta\delta_\delta$, and has no serious $\Delta\delta_\alpha$ term. The R.A. observations are likewise free from clamp effect, but show some unexpected anomalies in the dependence of SD on ZD.

Gill's transit-circle is now near the end of a very active career that began in 1903. It would not be practicable to remove this telescope to the Sutherland observing station of SAAO, now established in the Karroo, and it seems far more fitting that it should remain on its own site as an example of instrumental design of the beginning of the twentieth century that was several decades ahead of its competitors.

To take its place as part of the SAAO's astrometric effort, we hope to develop a transportable automatic transit circle that can be used at Herstmonceux and Sutherland to give fundamental positional coverage in both hemispheres.

DISCUSSION

Edmondson: As a non-expert I am shocked at the concept of a 'transportable' transit circle. Could you explain in more detail this departure from the traditional concept of leaving the instrument undisturbed for long periods of time?

Tucker: We are perhaps nowadays rather more confident in our ability to attain and control adequate instrumental rigidity. It is mainly a question of taking advantage of modern engineering techniques and materials.

Klock: In Washington we feel that the success of the stability of the 6-in. T.C. may be attributed to 2 important factors:

(1) The design and stability of the pier foundation for the instrument, and

(2) The stability of the azimuth marks.

Gliese, Murray, and Tucker, 'New Problems in Astrometry', 95. All Rights Reserved.

REPORT ON THE CAPE PHOTOGRAPHIC SURVEY

S. V. M. CLUBE

Royal Observatory Edinburgh, U.K.

and

W. NICHOLSON

Royal Greenwich Observatory, Herstmonceux, England

Since the last report (Clube, 1970), the Cape fourfold overlapping survey of the southern hemisphere has been completed. Approximately one half of the ~ 6000 plates have now been transferred to Herstmonceux for measurement. The measurement procedure is two staged: initially some 300 of the brightest stars on each of one non-overlapping set of these plates are picked out visually and paper tape listings of rough (x, y) co-ordinates are produced with a D-Mac machine. After some computer editing, these tapes are used to generate search tapes for measuring all the plates on a GALAXY machine. Some 1100 plates have now been measured this way, but the advance to GALAXY measurement has not yet been made on any significant scale. This is because of two difficulties which it is hoped to resolve in the near future, namely the existence of a small astrometric magnitude equation in GALAXY, and the presence on some of the plates of small glass splinters which can damage the 'nose cone' of the measurement optics as the plate is tracked beneath it. The technique of reduction and analysis is not described here, but it may be mentioned that the anticipated relative positional accuracy of ~0″.05 is being achieved over fairly wide areas of the sky.

DISCUSSION

Strand: It seems difficult to believe that a star position is obtained with a mean error of 0″.05, which means 0.5 μ on the plate, in view of the fact that the overlapping plates do not contain the same reference stars.
Clube: The typical external error of one 'plate-star' averaged from two exposures on each plate, after local smoothing of the system with the overlaps, is of the order of one micron. With each star appearing on at least four separate plates, the mean error approaches 0″.05.

Reference

Clube, S. V. M.: 1970, *Trans. IAU* **XIVA**, 43.

POTENTIALITIES OF YALE ASTROMETRIC MATERIALS

E. D. HOFFLEIT

Yale University Observatory, New Haven, Conn., U.S.A.

Abstract. A brief summary is given of the plate materials and results obtained at Yale since the advent of photographic astrometry. The distributions of plates not yet measured or reduced are indicated and possible supplementary programmes for their effective utilization are discussed. Interest centres chiefly in the following areas:
(1) The unmeasured parallax plates.
(2) Reduction of the $-60°$ to $-70°$ zone catalogue.
(3) The El Leoncito programmes for the determination of both positions and proper motions relative to faint galaxies.
(4) The feasibility of a search for intrinsically faint nearby stars with the Yale 40-in. at Cerro Tololo and the 20-in. double astrograph at El Leoncito.

The majority of photographic plates at the Yale Observatory stem from programs initiated by Frank Schlesinger (1871–1943, known as the 'Father of Modern Astrometry') and supplemented by the interests and energies of Dirk Brouwer (1902–1966). These comprise plates for stellar parallax, zone catalogues of positions and proper motions, as well as planet and asteroid programs planned for ascertaining corrections to star catalogues, and polar plates taken with the Loomis 15-in. polar telescope for investigating the motions of the Earth's axis. Under Dr Brouwer's direction several doctoral theses had been written based on the solar system observations (e.g. Pierce, 1971). To my knowledge no further investigations with these materials are contemplated. My present survey is limited to plates for parallax and proper motion programs.

1. Parallaxes with the 26-in. Refractor

Until W. L. Elkin's (1855–1935) retirement from the Directorship in 1910, Yale was among the last of the strongholds adhering to the classical visual technique (split objective heliometer) with which Bessel in 1838 determined the first accurate published parallax. Frank Schlesinger came to Yale in 1920 and the Observatory quickly became recognized as the foremost center for stellar parallax. The momentum of his achievements spurred his successor, Brouwer, with the devoted help of Louise Jenkins and Donald Kimball, to pursue stellar parallaxes right up to the present. The acquisition of parallax plates at Yale ceased in 1963 with the transfer of the 26-in. refractor to Mount Stromlo. After his recent retirement, Kimball has submitted his final list of some 20 parallaxes for publication. This marks the end of a long and distinguished photographic era during which Yale determined some 2100 parallaxes, mainly of southern stars.

However, the reductions of accumulated parallax plates are by no means complete. There remain nearly 6700 parallax plates taken between 1928 and 1963 still unmeasured. While many were intended for improving earlier results, there are available

Gliese, Murray, and Tucker, 'New Problems in Astrometry', 99–103. All Rights Reserved.

from 15 to 43 plates each on 59 stars for which no parallax at all has been determined, and on 53 for which only one determination at some other observatory is available. For nearly 300 stars additional plates had been taken for investigating variable proper motions. Many of the stars in these three groups are Luyten high proper motion stars. This constitutes potential material for mass ratios of nearby Population II stars for which astrophysicists have expressed an urgent need. Indeed, for 23 more stars with already well determined parallaxes, from 10 to 56 additional plates spanning 5 to 32 years had been taken for eventual mass ratio determinations. Although personnel and funding for computer time for this work are sadly lacking, the U.S. Naval Observatory has proffered help in permitting the use of its semi-automatic measuring engine for the plate measurements, and Dr L. Auer is preparing computer programs for their reduction.

The 26-in. refractor was also used for other programs, both astrometric and photometric. Among the former, as Dr Klemola has pointed out to me, are plates taken for the determination of proper motions of 22 RR Lyrae stars with median magnitudes between 9.5 and 14.0. From two to four plates each had been taken about 1928 and again in 1944–45 on the following stars: SW Aqr, S Ara, X Ari, RV Cap, TX Car, EE Car, AT Cen, RR Cet, RT Dor, RV Dor, RX Eri, SV Hya, RV Leo, V Lep, RZ Lib, T Men, V Men, ST Oph, RV Phe, SS Tau, VX Vel, and ST Vir.

2. The Zone Catalogues

Since 1920 the Yale Observatory has published catalogues of the positions and proper motions of approximately a quarter of a million stars mostly brighter than 11th visual magnitude between declinations $+85°$ to $+90°$, $+50°$ to $+60°$, $+30°$ to $-50°$, and $-70°$ to $-90°$ (the last only provisional). These results are published in most of the volumes of the *Transactions of the Yale University Observatory* from Vol. 3 through Vol. 31. All north of $-30°$ are largely the achievement Dr Ida Barney who had been Schlesinger's able assistant from the beginning and who carried on the work until her retirement in 1959. Only the zone from $-60°$ to $-70°$ has not yet been reduced. The plates for this zone had been measured under the direction of Dr Arnold Klemola in 1965. The former Army Map Service, who financed all of the Yale Zone Catalogue work for the zones south of $-30°$, had volunteered to carry out the reductions of this particular zone. All of Klemola's measurements as well as necessary auxiliary reduction data on punched cards were transmitted to the Army Map Service in Washington. However, they did very little work on this project, and last year all of the raw materials were returned to Yale. Reductions must now await adequate funding for programmers and computer time as well as for putting now warped cards onto tapes.

3. Proper Motions Relative to Faint Galaxies

This project for extending the Lick program to the South Pole had been initiated during the regime of the late Dirk Brouwer. It has become the province of Dr Wesselink who reports on the completion of the first epoch plates in his paper later in this Symposium. While there is by no means unanimous agreement on whether the second epoch plates should begin after a lapse of 10 or of 20 years, the problem is unfortunately resolved for the present by the lack of support of the Yale-Columbia Southern Station beyond 1973. This same lack of support militates as well against the initiation of other contemplated Yale projects for the 20-in. southern astrograph at El Leoncito.

4. Proposed Search for Faint Nearby Stars

Two recent independent investigations (Murray and Sanduleak, 1972; Weistrop, 1972) suggest that there are many more faint nearby red dwarfs than have already been found, or than are implied by the luminosity function derived by Luyten from his discoveries of high proper motion stars. The vast majority of the undiscovered stars would presumably be members of the disc population, with circular motions too nearly parallel to that of the sun to be detected in a selective survey for large proper motions. It is not my purpose here to discuss how valid is the evidence for their existence; rather I should like to consider the feasibility of a search for such stars with available Yale equipment should they indeed exist.

Several factors are considered: (1) The number of such stars to be expected in the areas of plates taken with the Yale telescopes; (2) The total number of stars in the field among which the parallax stars are to be detected; (3) The average number of high proper motion stars expected in the field; (4) The time factors necessary for making the search; (5) and most importantly, the potential accuracy of measurements for this purpose.

Table I summarizes the expected frequencies for plates reaching limiting visual magnitude 18. The numbers of faint red dwarfs within 20, 10 and 5 parsecs are based on Weistrop's luminosity function for the disc population. The number of high proper motion stars is inferred from Luyten's (1971) published results of his blinking 48-in. Palomar-Schmidt plates. The numbers of field stars are interpolated from the Sears-van Rhijn (1925) star counts on Mount Wilson plates.

TABLE I

Expected numbers of stars per sq deg

Category		No.
Parallaxes	$\geqslant 0\overset{''}{.}050$	1.2
Parallaxes	$\geqslant 0.100$	0.2
Parallaxes	$\geqslant 0.200$	0.02
Proper motions	$\geqslant 0\overset{''}{.}100$	20
Field stars at pole of Galaxy		660
Field stars at pole of Ecliptic		3400

Two telescopes might be available for the search. The southern El Leoncito 20-in. astrograph has a scale of $55''$ mm^{-1} and takes 17-in. square plates covering $6° \times 6°$. The Yale 40-in. reflector with a scale of $20''$ mm^{-1}, 5×7 in. plates covers 0.7 square degrees. It is now being re-located at Cerro Tololo. At least two parallax search plates would be taken at each of three consecutive epochs at intervals of approximately six months (the third in order to eliminate proper motion), at times of largest possible positive and negative parallax factors. If the plates are measured to an accuracy of one micron, the best accuracy that can be expected from a parallax determination is about $0''.05$ on El Leoncito plates, and $0''.02$ on plates taken with the 40-in. Thus the El Leoncito plates would be adequate for the detection of stars at best to 10 pc and probably not beyond 5 pc (one to 7 parallax stars per region photographed). The 40-in. could be of marginal value almost out to the desired 20 pc (30 to 40 parallax stars in an aggregate area equal to that of an El Leoncito plate.

With the Yale Mann measuring engine whose only automatic feature is to record measured coordinates on punch cards, the measurements of all of the stars on an El Leoncito plate reaching 18 mag. would require well over 900 h for the pole of the ecliptic and 200 at the galactic pole. The same total number of stars measured on the small plates taken with the 40-in. telescope would take longer because of more plate handling. With automatic or semi-automatic engines such as the GALAXY or the U.S. Naval Observatory machines, the measuring time would be reduced by approximately 75% and all of the stars in a 36 sq deg area at the galactic pole could be measured in little over a week. Moreover, with automatic image centering the setting errors might be reduced to half a micron.

If only one star in nearly 700 at the pole of the galaxy is expected to have a large parallax, one wishes these stars could be discovered by means of the blink comparator rather than through prolonged precision measurements. Luyten's experience during his Bruce Proper Motion Survey indicated that with plates having a scale of $60''$ mm^{-1} and a time span of 10 years he could detect stars with proper motions of $0''.100$, corresponding to image displacements of 0.02 mm between the two plates being blinked. This would imply that with the El Leoncito plates only stars within 2 pc could be found by this technique. On the 40-in. plates the limiting parallax discoverable would be close to $0''.20$, whence an average of 50 plates might have to be blinked to find just one nearby star – a futile effort for the purpose in hand.

Parallax stars indicated by precise coordinate measurements of a limited number of plates, as described above, should for the most part be considered only as good suspects whose parallaxes would ultimately have to be established by more extensive series of standard parallax observations.

References

Luyten, W. J.: 1971, *Proper Motion Survey with the Forty-Eight Inch Schmidt Telescope*, No. XXVII, Univ. Minnesota.
Murray, C. A. and Sanduleak, N.: 1972, *Monthly Notices Roy. Astron. Soc.* **157**, 275.
Pierce, D.: 1971, *Astron. J.* **76**, 177.

Sears, F. H., van Rhijn, J. P., Joyner, M. C., and Richmond, M. L.: 1925, *Astrophys. J.* **62**, 320.
Weistrop, D.: 1972, *Astron. J.* **77**, 366 and 849.

DISCUSSION

Luyten: I feel I must take issue with the last part of the paper: if there really is 1 star per sq deg nearer than 20 pc and brighter than 18^m, then nearer than 5pc (where they would be 3^m brighter) there should be 1000 stars brighter than 18^m and this I just can't accept.

As to Miss Weistrop's result: some twelve years ago I published an analysis made near the South Galactic Pole on three-image plates which Haro and I took with the Palomar 48-in. and I found a great dearth of M stars and of course faint stars nearer than 20 pc should be M dwarfs: but at that time this was an unpopular conclusion and it is now – so it was simply swept under the rug.

Murray: I don't want to be forced into a position about these stars. All I did was measure proper motions of 21 stars, and from these it appeared that the small dispersion and fairly large mean motion indicated that they were relatively nearby and had a small velocity dispersion. Arguments about the local mass density are based on gross extrapolation, and may well be wrong. If there are a large number of young objects around, then they may well appear in groups and perhaps in the north polar region we have hit such a group.

THOUGHTS ON THE FUTURE OF PHOTOGRAPHIC CATALOGUES

HARLEY WOOD

Sydney Observatory, New South Wales, Australia

Abstract. The current astrometric programmes of zone photography and minor planet observations at the Sydney Observatory are reviewed. Future instrumental requirements and resources needed for zone catalogue work are discussed.

At Sydney Observatory zone photography of the southern sky began with the object of repeating, with better means, our astrographic zone $-51°$ to $-65°$ and possibly the zone $-64°$ to the South Pole, originally allotted to Melbourne Observatory but largely completed and published from Sydney. The intention has been enlarged to embrace the whole southern sky. The camera is based on a Taylor, Taylor and Hobson lens of focal length 1776.6 mm (scale $116''$ mm^{-1}) and there is no deterioration of image quality over a field $6° \times 6°$. On a plate 20 cm × 20 cm the measurable area is over $5° \times 5°$ so that taking zone centres $2°.5$ apart in declination and with right ascensions appropriately spaced there is full overlap in both coordinates. All exposures are made through a coarse diffraction grating.

So far all of the zones centred at declinations from $-43°.5$ to $-63°.5$ have been completely photographed and $-41°$ and $-66°$ are virtually complete, and in fact would have been so except for some rejections. Zones at $-38°.5$ and $-68°.5$ have been commenced and will go ahead immediately after the IAU Assembly.

Results of observations of selected minor planets give data relevant to catalogues of star places. At Sydney we have made observations over long arcs of such minor planets since 1955 and I quote from some results compiled by W. H. Robertson. Each plate taken has four exposures separated in declination. The results are always subjected to comparison of two kinds. First the means for all four images are taken with two separate groups of reference stars and second the mean result, using all reference stars simultaneously, from the first two exposures is compared with the result from the last two exposures. Each comparison implies a standard error for the result from the whole plate (all images and all stars). The first way has continuously through the whole period given standard deviation $\pm 0''.23$ and the second way $\pm 0''.09$. Of course the second method eliminates completely the effect of errors in the star places and largely the effects of emulsion shift, metric properties of the image field and of errors in the measuring machine but not others, chiefly pointing errors. We have always tended to regard the difference between the two results as due to defects in the proper motions used to bring the star places to the epoch of the plate.

In 1955 and 1956, initiating during a visit by Dirk Brouwer, the plates on which a series of Yale Catalogues (e.g. Hoffleit, 1967) are based were taken at Sydney Observatory. The publication of these catalogues enables the plates in the appropriate area

Gliese, Murray, and Tucker, 'New Problems in Astrometry', 105–108. All Rights Reserved.

to be reduced again with contemporary star places. This does not affect the second class of comparison but the comparison using two sets of reference stars separately now implies a standard deviation $\pm 0\rlap{.}''09$.

Recently we received from V. Orelskaya of the Institute of Theoretical Astronomy ephemerides of selected minor planets depending on observations collected from many sources. The deviations $(O-C)$ of our results from these ephemerides make possible another estimate of errors. The deviations usually keep nearly the same value over a long arc due possibly to a system in the catalogue from which the reference stars came or to a need for improvement of the orbit. A correction for this systematic difference having been applied the deviations from the ephemeris imply a standard error in each coordinate $\pm 0\rlap{.}''23$ for the original results and $\pm 0\rlap{.}''07$ for results using the contemporary star places.

The defects in the proper motions are verified and in fact a new reduction using the older catalogues but neglecting proper motion entirely leads in about 40% of cases to better agreement between results on one plate from the two sets of reference stars. These observations are made in areas of the sky not usually regarded as having the worst places and the need for continued positional work of quality is underlined.

It is scarcely necessary to enlarge on, or even to mention, *to* a group of astrometrists the fundamental part played by positional astronomy in many general activities of astronomy and astrophysics. It provides for the identification and recognition of many objects, for the observations needed in celestial mechanics including space science. Proper motions, obtained by positional techniques, provide a sieve to identify objects of various classes, for example those belonging to a cluster or association or those near to us in space, and form a part of the basic data needed in galactic kinematics. However there is plenty of evidence that it is necessary that it should be said with emphasis *by* the positional astronomers. At present there are at least three large telescopes, each costing perhaps 15 million dollars, under construction for use in the Southern Hemisphere. Much more is of course offered to space science. Half the cost of one such telescope devoted to positional astronomy would make a very great contribution to astronomy as a whole.

The same care and effort in design and choice of site (and comparable resources for construction) devoted to a great telescope needs to be available for positional projects. Due to the devotion of some positional astronomers progress on the design side is quite appreciable as revealed in reports of development of the Automatic Transit Circle at Washington and in the draft reports prepared by Fricke and Vasilevskis as presidents of IAU Commissions 8 and 24. This would doubtless be greater if there were assured prospects for construction and siting for most efficient use of the additional resources. The question as to the direction in which progress should be sought needs to be raised.

Leaving aside the means of solution (see for example Fricke, 1972) the problems of optical astrometry in fact remain much the same; to increase the accuracy of positions and proper motions and, within the best standard of accuracy, to establish a good self-consistent fundamental system over the whole sky and to relate through good proper motions observations both past and present to the same system which can be carried into the future. In some areas the relation between the fundamental system and the photographic zone catalogues (Fricke, 1972) remains weak. It seems important that the problem should be considered on a whole sky, all magnitude basis.

The availability of overlapping techniques which require fewer reference stars (for example Eichhorn and Gatewood, 1967) for reduction of measurements for photographic catalogues can work in two directions. They should mean on the one hand that effort devoted to meridian programmes might be directed less to providing a mass of observations of many stars and more to concentrating on the fundamental character of the observations and on the other hand that smaller fields, and longer focal lengths, can be used in the photography.

Longer focal lengths for this kind of work, (although suggested more than once), have not been much exploited, except in restricted areas (Eichhorn *et al.*, 1970) where the success has been conspicuous. Their use would lead to greater accuracy through reduction of the influence of particularly measuring errors and emulsion shift. To estimate this in the conditions in which catalogues work is done, with measurements made over the whole plate, the model for the emulsion shifts needs to be carefully considered and the experience of Eichhorn and his collaborators who obtained a relative accuracy of 'the order of $0''.01$', supports the use of larger focal lengths. Part of the accuracy was, of course, due to redundancy of material and advanced methods of reduction but it could not have been produced by the use of plate material at a scale of about $100'' \text{ mm}^{-1}$ such as is commonly used for catalogue work.

Then what sort of instrument should be specified? Suppose that the largest plate size that can be tolerated is 40 cm × 40 cm and the smallest field 4° × 4°. Allowing for an unmeasured border around the plate edge this leads to a focal length of over 500 cm and a scale $40'' \text{ mm}^{-1}$, greater by a factor of 2.5 over frequently used scales.

Then the optical system needs to be considered. Although there are differing views on this, it does seem that a refracting system is to be desired. Luyten who has had such admirable success in the use of a Schmidt camera for measurement of proper motion nevertheless says (Luyten and La Bonte, 1972) that 'absolute' positions are "certainly not as good as can be obtained from Carte-du-Ciel plates (even allowing for the slight difference of scale)" and Dieckvoss (1972) speaking of the solution for positions on whole Schmidt plates says that if an observer "has to use Schmidt plates for photographic astrometry he has to restrict himself to rather small areas". In cases where experiments have been made with Schmidt cameras one usually finds that the results are not competitive or the area used has been restricted. There can be no doubt of the very real application of the Schmidt camera in the latter case.

However it is most important that there should be no compromise with the quality of the optical system. The images must be good over the whole field which must have excellent metric properties. The difficulty of arranging this depends on the magnitude limit sought but an aperture of less than 40 cm should be adequate. With the consequent aperture ratio of about $f/13$ the lens designer should have a quite tractable problem. Estimates have been made of the astrometric accuracy as a function of scale but all that it seems necessary to say is that the standard error should for the first time come down to about 0".05.

In the past the measurement has been the barrier to quick progress in forming photographic catalogues. This need no longer be the case since measuring rates of 500 stars per hour (Reddish, 1972) are reported and even higher (Kibblewhite, 1972) projected. With a new machine for astrometry to a magnitude limit of say 12 and able to measure an area 40 cm it ought to be quite possible to keep up with the output of two cameras, one in each hemisphere.

The design of the telescope mounting and accessories too needs to provide for efficient working. It should be quite possible to arrange for all settings for each plate centre to be available in machine readable form and made by the turn of a switch and for the guide star to be found and followed automatically. The observer would then merely need to load and unload the plate holders. With two such a telescopes set up at well selected sites it would be possible to cover the sky in a period short compared with what we have been used to in the past.

References

Dieckvoss, W.: 1972, *Conf. on the Role of Schmidt Telescopes in Astronomy*, Hamburger Sternwarte, p. 43.
Eichhorn, H. and Gatewood, G. D.: 1967, *Astron. J.* **72**, 1191.
Eichhorn, H., Googe, W. D., Lukac, C. F., and Murphy, J. K.: 1970, *Mem. Roy. Astron. Soc.* **73**, 125.
Fricke, W.: 1972, *Ann. Rev. Astron. Astrophys.* **10**, 101.
Hoffleit, D.: 1967, *Trans. Astron. Obs. Yale Univ.* **28**.
Kibblewhite, E. J.: 1972, *Observatory* **92**, 221.
Luyten, W. J. and La Bonte, A. E.: 1972, *Conf. on the Role of Schmidt Telescopes in Astronomy*, Hamburger Sternwarte, p. 35.
Reddish, V. C.: 1972, *Observatory* **92**, 220.

DISCUSSION

Strand: I wonder if you have overlooked the high cost of building a machine measuring 40 cm × 40 cm. I estimate it to cost between $ 500 K and $ 750 K, if not more.

Wood: No. This could well be among the items provided from the resources taken to be equivalent to half the cost of a large reflector.

Not all of these resources are allotted in the paper. What about adding two meridian circles to the two cameras and one measuring machine mentioned? This would leave enough for site selection and development and if there were a surplus it could be spent on an astrometric reflector. I find it hard to imagine expenditure on astronomy which would give better value per dollar.

Eichhorn: I agree that doing zone catalogue work with a telescope that gives a scale of 40" mm^{-1} would give a much better return in accuracy per effort than the present 100" mm^{-1} which was first used by Schlesinger and has since become almost a 'magic number'. Increasing the focal length is the easiest way to achieve higher positional accuracy, and it is remarkable that this has not been used more frequently.

Vasilevskis: Would it not be better to use the astrographs already at Lick and El Leoncito for cataloguing work? The scale is favourable at 55" mm^{-1}.

Wood: Well, I did think of this but, by a small margin, decided not to mention it. Perhaps I was wrong. If the telescopes were available for the necessary series of plates, close consideration would need to be given to the question.

Eichhorn: In answer to Vasilevskis' remark, $55^2 = 3025$, and $40^2 = 1600$, so that the weights of a position obtained by telescopes with scales of 40" mm^{-1} and 55" mm^{-1} would be approximately as 2:1.

SESSION C

RADIO ASTROMETRY

RADIO ASTROMETRY USING CONNECTED-ELEMENT INTERFEROMETERS

(Invited Paper)

B. ELSMORE

Mullard Radio Astronomy Observatory, Cambridge, England

Abstract. The basic techniques of conventional radio interferometry are explained and a review given of the achievements and limitations of the methods as applied to radio astrometry.

The application of radio astronomy techniques to establish an astrometric system based on extragalactic radio sources has important consequences both in radio astronomy and in optical astrometry. In radio astronomy it will provide a reference frame of high accuracy for source catalogues and for the calibration of instruments for positional measurements. In optical astrometry, a comparison between the positions of compact, extragalactic objects determined by radio and optical methods may lead to a reduction in the errors in fundamental optical catalogues that arise from effects due to proper motion of individual stars and of galactic rotation, and hence, to the elimination of large scale inhomogeneities (Fricke, 1972).

Unlike the situation in optical astronomy, no radio astronomy instrument has yet been designed primarily for astrometric purposes but two distinct interferometer techniques exist that provide sufficient precision in measuring the positions of radio sources to merit the use of the term 'astrometric'.

One method, that uses interferometers of very large baselines, sometimes as much as a few thousand kilometres, has necessitated the development of a timing and recording technique to enable the signals to be recorded separately at each end of the baseline and compared at a later time. The accuracy of this VLBI system is potentially extremely high, but it has not yet been fully realized in practice; one serious limitation being caused by the difficulty in knowing precisely the delay due to the ionosphere and atmosphere above each of the two widely separated aerials. In addition to astrometric uses, VLBI measurements have important applications in geodesy.

Another interferometric technique, which is my main concern here, uses a baseline of a few kilometres, so that the aerials may be linked by cables to a receiving system that instantly measures and records the difference in phase between the signals arriving at the aerials. From measurements of this difference of phase at two aerials, fixed on the surface of the rotating earth, the positions of radio sources may be derived. This conventional type of interferometer has been widely used in recent years and an estimate of the accuracies that have been achieved may be obtained from an examination of Table I. It should be noticed that none of these measurements has been made in the southern hemisphere.

Gliese, Murray, and Tucker, 'New Problems in Astrometry', 111–117. All Rights Reserved.

As the methods used in radio astronomy are very different from those of classical optical astrometry, an attempt will be made to give a broad outline of the principles involved.

The method of relating the measured phase difference to the position of a source on the sky clearly depends upon the orientation of the interferometer baseline and

TABLE I

Highest accuracies claimed for various instruments

			"(arc)
NRAO Greenbank	Wade	(1970)	0.6
Cambridge One Mile	Smith	(1971)	0.2
RRE Malvern	Adgie et al.	(1972)	0.4
NRAO Greenbank	Brosche et al.	(1973)	0.1
Cambridge 5 km	Ryle and Elsmore	(1973)	0.02

the geometry involved is perhaps the simplest when the interferometer baseline lies east-west. In the early days of radio interferometry, when the aerials were large fixed arrays, unable to track radio source, observations were made only at transit and the east-west interferometer was then analogous to the meridian circle, being subject to the usual level, azimuth and collimation errors. Now that much shorter radio wavelengths are used, and aerials track the sources as the earth rotates, the R.A. and Decl. may be determined independently.

If a vector represents the baseline spacing between two aerials on the Earth's surface as shown in Figure 1a, it can be seen that the vector, due to the assumed uniform

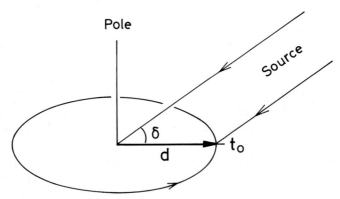

Fig. 1a.   A vector representing the baseline spacing of an east-west interferometer is shown to sweep out a circle due to the rotation of the Earth. The measured phase difference at any instant is proportional to the component of this vector in the line of sight from the source.

rotation of the earth, rotates in a plane at right angles to the polar axis; the sideways motion is of no consequence as it only produces an aberration. The path difference at any instant is the component of the vector in the line of sight, and hence, the measured phase difference is

$$\phi(t) = \frac{2\pi}{\lambda} d \cos(t - t_0) \cos\delta + C$$

where C is the electrical collimation error; a delay within the receiving system caused, for example, by unequal cables connecting the aerials to the receiver. If we now plot the measured ϕ against time we obtain a curve of the form shown in Figure 1b.

Declination may be derived from the amplitude of the sinusoidally varying component, provided that the spacing d is known. A disadvantage can be seen of the east-west interferometer system in that the accuracy of measuring declination

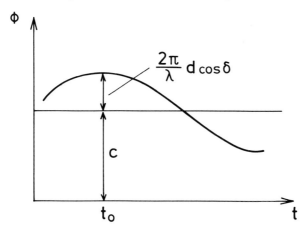

Fig. 1b.   The measured phase difference, plotted against time, for an east-west interferometer system.

varies with $\sin\delta$. However, with the 5 km telescope at Cambridge (Ryle, 1972) this fact has been put to an advantage and used to determine the aerial spacing independently of the initial land survey of the instrument, by observing radio sources of only approximately known declinations lying near the equator. Thus the distance, for example, between two of the aerials has been measured to be $3\,430\,828.7 \pm 0.25$ mm (i.e. to 1 in 10^7 which is about 10 times more accurate than that available for the measurement of 3.4 km by conventional means). Using this value for spacing, declinations of other sources may then be determined absolutely.

Hour angle may be derived from the phase of the $\phi(t)$ plot, but the relation between hour angle and right ascension presents more difficulty. The lack of suitable bright and compact radio objects in the solar system makes it difficult to establish the ecliptic and hence the equinox from radio observations, and so for the 5 km telescope the zero point of R.A. has been established from observations of β Persei (Algol) which intermittently radiates sufficiently strongly at cm wavelengths, thereby enabling the RA scale to be related directly to FK4 (Ryle and Elsmore, 1973). A redetermination of the optical position of β Persei at the Royal Greenwich Observatory and at the Institute of Astronomy, Cambridge has shown the FK4 right ascension to be in error by less than 3 ms (Tucker et al., 1973).

With an interferometer not aligned east-west, the baseline vector sweeps out a cone as the Earth rotates as shown in Figure 2a. This fact has been utilised ingeniously by Wade (Wade, 1970; Brosche *et al.*, 1973) of NRAO to determine declinations without prior knowledge of the baseline or of any source declinations. As before, the phase difference is $2\pi/\lambda$ times the component of the vector in the line of sight. As B

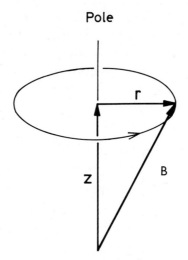

Fig. 2a.   The baseline vector for a non east-west system describes a cone as the Earth rotates. The vector may be resolved into a component z, parallel to the polar axis, and a rotating component r.

describes a circle, the phase varies sinusoidally, hence

$$\phi(t) = \frac{2\pi}{\lambda}\left[z \sin\delta + r \cos\delta \cos(t - t_0)\right] + C.$$

Three measured parameters may be derived from a plot of $\phi(t)$, as shown in Figure 2b.

(1) A constant term, $\phi_c = (2\pi/\lambda) z \sin\delta + C$.

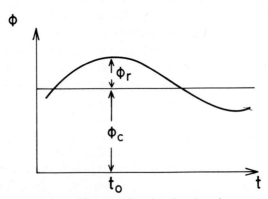

Fig. 2b.   The measured phase difference, plotted against time, for a non east-west system.

(2) The amplitude of the oscillatory term, $\phi_r = (2\pi/\lambda)\, r \cos \delta$.

(3) The time t_0, which is related to hour angle.

For several sources, say N, with δ_i, z, r and C unknown. Solving for declination:

> $N + 3$ unknowns: δ_i, z, r and C
>
> $2N$ equations from (1) and (2) (i.e., from ϕ_c and ϕ_r).

Therefore, a solution may be obtained for declinations using observations of three sources, and furthermore, the method remains accurate at low declinations. Relative right ascensions may be derived from the phase of the $\phi(t)$ plot and at NRAO, the zero of R.A. has been established for their instrument from observations of four extragalactic radio sources for which there are accurate optical positions, (Brosche *et al.*, 1973).

On careful examination of the properties of these two differently orientated interferometers, it can be seen that there is another difference, in that for the east-west system, the measurement of declination and right ascension is independent of the instantaneous position of the pole.

Consider an east-west interferometer situated on the Greenwich meridian. The x component of polar motion changes the latitude, which has no effect on the phase difference between the two ends of the baseline. The y component will rotate the baseline to give a small displacement parallel to the mean polar direction that contributes to C, the electrical collimation term, but will, to first order, make no change in the component parallel to the equatorial plane. Therefore, the measurement of R.A. and Decl. remains unchanged. The polar motion that occurs during the observations is sufficiently small as to not affect the measurements. For an interferometer not at Greenwich a similar argument holds, since the instantaneous polar coordinates can be resolved in, and at right angles to the local meridian.

This is not the case for a non east-west interferometer that has an appreciable component of the baseline vector parallel to the polar direction.

An advantage over all other methods of the technique that utilize moderately spaced interferometers stems from the fact that only *differences* between the paths to the two aerials are involved, hence atmospheric corrections are only of second order and are typically $1''$. With a horizontally stratified atmosphere the path differences, and hence the corrections, would be zero, although the individual aerials would have to be steered to allow for refraction. However, in practice, path differences occur due to the curvature of the atmosphere and ionosphere, for which correction is possible. Although the aerials of an interferometer may only be a few kms apart, irregularities in the water vapour content of the troposphere produce phase variations for which no correction is possible. These effects, which are most severe during the daytime in summer, are the dominant factor that limit the accuracy that can be achieved by the 5 km telescope (Ryle and Elsmore, 1973). The magnitude of effects arising from various external causes for different interferometer spacings are shown in Table II (from Hinder and Ryle, 1971).

What of the future? The 5 km telescope at Cambridge has some features that make

TABLE II

Uncertainties of differential path in mm at 30° elevation for Cambridge

Baseline	Troposphere irregularities		Troposphere curvature	Ionosphere irregularities at 5 GHz
(km)	Summer	Winter		
1	2.6	0.66	0.12	0.03
10	2.8	0.71	1.2	0.35
100	2.8	0.71	12	2.8

Note: 1 mm at 1 km = 0″.2.

it especially suitable for astrometry, (for example, the declination and hour angle axes intersect at a point), but the instrument has been designed primarily as a fully automated instrument for providing high resolution maps of radio sources. For this reason, four of the eight aerials are mobile and are mounted on rails which renders them less stable, and so for astrometric observations, only the fixed aerials are used, giving spacings up to 3.4 km. From preliminary observations of much of the 3C catalogue north of Decl. 10°, only 12 sources have so far been found to be sufficiently compact for the highest precision astrometric purposes; all the other sources examined are either extended or complex. The positions of these 12 sources, with an accuracy of about ±0″.03 are about to be published (Ryle and Elsmore, 1973). It is estimated that there are another 100 suitable sources in the N hemisphere brighter than 0.2 f.u. In this connection, it is noted that a working party of Commission 40 has been set up to compile a list of sources suitable for positional calibration purposes.

It is particularly desirable that sources at low declinations should be observed so that the positions of sources in the N hemisphere may be linked to those which it is hoped will be measured in the S hemisphere. This cannot be done with east-west interferometer systems, such as those at Cambridge and Westerbork.

The contribution to astrometry from connected-element radio interferometry is only just beginning, now that large angles on the sky can be measured with great precision and that the positions of radio sources may be determined with comparable, if not greater accuracy than those of optical observations of stars. It is very important, in addition, to appreciate that the factors that limit the accuracies fortunately have different effects in radio and in optical astrometry, thus making the two types of observations complementary.

References

Adgie, R. L., Crowther, J. H., and Gent, H.: 1972, *Monthly Notices Roy. Astron. Soc.* **159**, 233.
Brosche, P., Wade, C. M., and Hjellming, R. M.: 1973, *Astrophys. J.* **183**, 805.
Fricke, W.: 1972, *Ann. Rev. Astron. Astrophys.* **10**, 101.
Hinder, R. A. and Ryle, M.: 1971, *Monthly Notices Roy. Astron. Soc.* **154**, 229.
Ryle, M.: 1972, *Nature* **239**, 435.
Ryle, M. and Elsmore, B.: 1973, *Monthly Notices Roy. Astron. Soc.* **164**, 223.

Smith, J. W.: 1971, *Nature Phys. Sci.* **232**, 150.
Tucker, R. H., Yallop, B. D., Argue, A. N., Kenworthy, C. M., Ryle, M., and Elsmore, B.: 1973, *Monthly Notices Roy. Astron. Soc.* **164**, 27P.
Wade, C. M.: 1970, *Astrophys. J.* **162**, 381.

DISCUSSION

Tucker: Does the solution for three sources need three distinct δ values? Also there is a diurnal term in latitude variation that amounts to $0''006$.

Elsmore: The method does, in fact, rely upon the use of three different declinations. Concerning the second point, I am grateful to Mr Tucker for making me aware of this effect and one must consider what result it will have on the positional measurements.

Eichhorn: Will it sometimes be possible to use the radio emissions from Jupiter and Saturn to establish the position of the vernal equinox – should this be deemed desirable – or are their diameters too large for this purpose?

Elsmore: The difficulty here is that the sources are extended and furthermore the centre of radio emission may not coincide with that of the figure of these planets.

Fricke: Could our colleagues from Greenwich or Cambridge, England, tell us how they succeeded in checking the right ascension of β Persei?

Murray: Algol was observed photographically at Herstmonceux and Cambridge relative to AGK3 stars, and also on the Herstmonceux Transit Circle relative to about fifteen FK4 stars on 20 different nights between September and December, 1972.

Teleki: We must compare, very carefully, the optical and radio observations. I suppose that there are many problems in this comparison including refractional problems. For this reason Prof. Fricke proposed to me, last year, the inclusion of a radio-astronomer in the Study Group on Astronomical Refraction. Dr W. J. Altenhoff (Max-Planck-Institut für Radioastronomie, Bonn), radioastronomer, has accordingly become a member of this Study Group. In his report prepared for the Study Group, I find this conclusion:

"In my feeling the astronomical refraction is not fully understood in the radio range. For single dish investigations one could think of some differential methods to measure the total refraction (with a dual beam system of fixed separation), the ionospheric refraction (with simultaneous observations at two frequencies) etc. Even though these measurements might help us to understand the refraction, it is not clear if they could help to improve positional (interferometric) work in radio astronomy."

Do you agree with this conclusion?

Elsmore: In interferometric work, the atmosphere does not present such a serious problem as you suggest, as only the difference between the paths to the two aerials is involved, and hence corrections are only of second order.

Murray: Could Mr Elsmore say something about the effect of the orbital motion of Algol? I believe you detected it at Cambridge.

Elsmore: Although the radio observations do not yet cover a complete period of the $1\overset{d}{.}9$ orbit, it is clear that the emission is almost certainly related to the *AB* system rather than to component *C*, which is separated from *AB* by $0''1$.

POSITIONS OF EXTRAGALACTIC RADIO SOURCES
FROM VERY LONG BASELINE INTERFEROMETRY

(Invited Paper)

C. C. COUNSELMAN III

Massachusetts Institute of Technology, Cambridge, Mass. 02139, U.S.A.

Abstract. The positions of extragalactic radio sources have been determined by very-long-baseline interferometry with uncertainties smaller than $0\overset{''}{.}1$, and fundamental limits on the accuracy of VLBI astrometry have not yet been reached. Technical improvements within the next five years may enable positions
to be determined by VLBI with uncertainties of $0\overset{''}{.}005$.

1. Introduction

Extragalactic objects can provide an excellent approximation to an inertial reference
frame. Thus it is important to determine their positions as accurately as possible. By
very-long-baseline interferometry (VLBI), the positions of extragalactic radio sources
can now be determined with uncertainties smaller than $0\overset{''}{.}1$, as demonstrated by the
consistency between results obtained by different groups of observers who used
different very-long-baseline interferometers and significantly different techniques.

2. VLBI Techniques

VLBI requires no real-time communication or connection between the separate
antennas which form the interferometer.* The signal received at each antenna is
converted to a low frequency and recorded on tape with a time base provided by the
local station clock. Tapes which have been recorded at separate antennas simultaneously observing the same radio source are later brought together and played back
to determine the difference between the time bases, or readings of the independent
station clocks, for which the crosscorrelation between the recorded signals is greatest.
Except for small instrumental and atmospheric delays, aberration, and random noise,
the resulting value of this time-difference is given by

$$\Delta\tau = \frac{1}{c}\mathbf{B}\cdot\hat{\mathbf{s}} + \Delta\tau_{\text{clock}}$$

where c is the speed of light; $\mathbf{B}$ is the vector position of one antenna with respect
to the other, known as the interferometer 'baseline' vector; $\hat{\mathbf{s}}$ is a unit vector in the
direction of the source; and $\Delta\tau_{\text{clock}}$ stands for the error of synchronization between
the two station clocks.

---

* This feature, rather than the physical distance between the antennas, distinguishes VLBI from conventional, connected-element interferometry.

Gliese, Murray, and Tucker, 'New Problems in Astrometry', 119–124. All Rights Reserved.
Copyright © 1974 by the IAU.

If the baseline vector **B** and the clock offset $\Delta\tau_{\text{clock}}$ could be determined independently, then an unknown source-direction unit-vector $\hat{s}$ could be determined by making only two observations several hours apart, using the Earth's rotation to obtain different projections of **B** upon $\hat{s}$. (Note that the baseline vector should not be nearly parallel to the Earth's spin axis. Neither should it lie too close to the equatorial plane; if it did, the accuracy of determinations of the declinations of near-equatorial sources would be degraded.) It is possible to determine simultaneously all 3 components of an unknown baseline vector, the clock offset $\Delta\tau_{\text{clock}}$ as a continuous function of time, and both coordinates of all sources observed by a 2-station interferometer if at least 3 well-distributed sources are observed for several hours. In this determination an origin of longitude or right ascension must be defined by some external means. Usually the right ascension of one of the sources, or the mean of several, is fixed. However, declinations are determined absolutely with respect to the equator of date.

The discussion up to this point has been appropriate for VLBI observations using wide effective bandwidth, for which accurate and unambiguous determination can be made of the *group-delay* difference between the signals received at the two antennas. VLBI observations can also be done usefully with narrow bandwidth, such that only the time-variation of the *phase-delay* difference (sometimes called the 'fringe phase'; the time-derivative of fringe phase is often called the 'fringe rate') can be determined with significant accuracy. With only phase-delay or rate observations it is impossible to determine $\Delta\tau_{\text{clock}}$ and the component of **B** parallel to the Earth's spin axis, and it is impossible to determine the declinations of all of the sources simultaneously with the length of the equatorial projection of **B**. The sensitivity of these observables to the declination δ of a source varies as $\sin\delta$, approaching zero at the equator. Usually *a priori* information is used to constrain the declination of a near-equatorial source, so that the declinations of the remaining sources and the equatorial projection of **B** can be determined from narrow-band VLBI data.

A general introduction to VLBI by Cohen (1973), a description of a VLBI tape-recorder system by Clark (1973), a mathematical analysis of the tape-processing by Moran (1973), and other related papers are collected in the September 1973 Special Issue on Radio and Radar Astronomy of the *Proceedings of the IEEE*.

3. Accuracy: Present and Future

To assess the present accuracy of source-position determinations by VLBI, I have compared the results obtained independently by two different groups of observers who used different very-long-baseline interferometers and significantly different techniques which are summarized in Table I. One set of results was obtained by T. A. Clark of the Goddard Space Flight Center of the U.S. National Aeronautics and Space Administration, A. E. E. Rogers and H. F. Hinteregger of Haystack Observatory, and C. C. Counselman III, C. A. Knight, D. S. Robertson, I. I. Shapiro, and A. R. Whitney of the Massachusetts Institute of Technology; they were first reported

TABLE I

Parameters of two independent determinations of radio source positions by VLBI. See text for explanations

Participating Organizations	GSFC, Haystack, MIT	JPL, INTA
Stations used	Haystack, Goldstone, NRAO, OVRO[a]	Goldstone, Madrid
Baseline length (km)	3900[b]	8400
Wavelength (cm)	4	13
Observable	Group delay	Fringe rate

[a] The National Radio Astronomy Observatory in Green Bank, West Virginia and the Owens Valley Radio Observatory in California were used only in 1969.
[b] Haystack-Goldstone.

by Clark *et al.* (1973), and are presented in detail by Rogers *et al.* (1973). These results were based on four separate periods of observation in 1969 and 1972, using mainly the 37-m diameter antenna of the Haystack Observatory in Tyngsboro, Massachusetts and the 64-m diameter antenna of the NASA Deep Space Net station in Goldstone, California. The observations were of group delay, for bandwidths of 35 MHz in 1969 and 23 MHz in 1972, centered on a frequency near 8 GHz ($\lambda \approx 4$ cm).

These source-position results are compared to the preliminary positions reported at this Symposium by J. L. Fanselow, P. F. MacDoran, J. D. Thomas, D. J. Spitzmesser, L. Skjerve, and J. G. Williams of the Jet Propulsion Laboratory, and J. Urech of the Instituto Nacional de Technica Aeroespacial, Madrid, Spain. There are 8 sources whose positions have been determined by both groups: 3C84, 3C120, OJ287, 4C39.25, 3C345, PKS2134+00, VRO42.22.01, and 3C454.3. For these 8 sources the rms difference between the right ascension determinations of the two groups was 0.005, or 0".08. Excluding the low-declination sources 3C120 and PKS2134+00 from comparison (because the fringe-rate observable used by the JPL-INTA group is insensitive to declination at the equator), the rms difference between the declination determinations of the two groups is 0".07. It is perhaps worth noting that the rms disagreement between the two groups' results is actually smaller than the published uncertainties; since the two groups worked completely independently this may signify that their quoted uncertainties are pessimistic.

What determines these uncertainties, and what improvements can we expect within a few years? The main sources of error are (i) the atmosphere, (ii) instrumental drifts, and (iii) geophysical-astronomical model inadequacies. I will discuss these in turn:

(i) *Atmosphere*. The equivalent electrical path length of the earth's neutral atmosphere, at 30° elevation, is about 5 m, but the *a priori* uncertainty is probably less than 60 cm. Assuming that the variations of atmospheric path length at opposite ends of the interferometer baseline are uncorrelated (any positive correlation only reduces the effect on the interferometric delay observable), we find that the equivalent source position uncertainty for a 4000-km baseline is about 0″.04. This uncertainty may be reduced significantly by modeling and making use of observations taken over a wide range of elevations to determine parameters in a model of the atmosphere simultaneously with the source coordinates, baseline vector components, and clock-synchronization parameters. However, a new method (Schaper *et al.*, 1970) involving microwave radiometric measurements of the emission of atmospheric water vapor (the constituent responsible for most of the path-length uncertainty) appears capable of reducing the uncertainty in atmospheric path-length at 30° to less than 3 cm, which corresponds to an uncertainty in source position of only 0″.002.

Depending on the radio frequency used in VLBI observations, the ionosphere may constitute a significant error source. At the 8-GHz frequency of the Haystack-Goldstone observations, the uncertainty introduced by the ionosphere is on the order of 0″.005, but this uncertainty could easily be reduced to a negligible level by using a higher frequency (e.g., several antennas exist which are suitable for VLBI observations at 15 GHz) or by making simultaneous observations at a lower frequency and using the f^{-2} dependence of ionospheric delay to calibrate this delay.

(ii) *Instrumental drifts*. Until recently, relatively little attention has been paid to measuring and correcting for variations in instrumental delays due to varying temperature, etc., in VLBI equipment. (Delay calibration of short-baseline instrumentation has received much more attention, because the effects of instrumental delay on source-position results increase with decreasing baseline length.) However, delay-calibration equipment suitable for VLBI has now been developed (by A. E. E. Rogers at Haystack Observatory) and installed at several antennas. With this equipment, drifts on the order of 1 ns in delay, equivalent to 0″.015 in position for a 4000–km baseline, have been measured, and should now be removable within less than 100 ps or 0″.0015.

(iii) *Geophysical-astronomical model*. In this category I include the problem of calculating the relative motions of the antennas due to solid-Earth tides, and the rotation of the earth. The effects of tides are clearly evident in Haystack-Goldstone VLBI results: a significant reduction in the post-fit residuals was obtained when a simple tidal model was included in the data-reduction program for the first time. Crustal motions on a time scale much longer than a day do not interfere with the determination of souce positions because the baseline can always be re-determined separately on each day of observation. It is not yet clear what problems remain to be solved in the area of Earth rotation, but the ability of group-delay VLBI to determine *both* the source positions and the vector baseline on a time scale of hours will be important as these problems are sorted out in the future. If a set of extragalactic sources is used arbitrarily to *define* a reference frame, then the rotation of the baseline vector with respect to this frame may be determined, simultaneously with clock-synchronization error, as quickly as the antennas can be moved through a well-distributed set of 4 sources. It seems likely that VLBI will play an important part in future determinations and improvements of theoretical models of precession, nutation, polar motion, and the variation of UT1.

4. Recent Results

Since the results compared in the preceding section were based on observations made in some cases as much as 4 years ago, one might ask what has been achieved in more recent, presumably more accurate, work. Preliminary results have been obtained from a series of VLBI observations made by the GSFC-Haystack-MIT VLBI group in October, 1972, using 2 antennas at Haystack Observatory and 2 antennas at NRAO, Green Bank, in a differential-interferometric mode. In this mode, a pair of sources is observed simultaneously using the same station clock for both sources at each end of the long baseline. The *relative* positions of the sources may thus be determined free from effects of clock instability. Using this data, Knight (1973) has made an independent determination for each of 8 days of observation, spanning 4 weeks and under very different weather conditions, of the position of quasar 3C 279 relative to another quasar, 3C 273.* The rms scatter of these 8 independent determinations of differential right ascension was only 0".02. Declination scatter was much greater, partly because only fringe phase observations were used and 3C 279 is only 5° from the equator: the rms was 0".08. Significant systematic errors which would not be revealed by the day-to-day scatter of position determinations may, of course, be present. Comparisons of these with other recent results will be necessary before the true uncertainties can be estimated.

5. Conclusion

It seems reasonable to expect that within five years the uncertainties of radio source position determinations by VLBI will be as small as 0".005. However, this low level of uncertainty will probably apply only to the *relative* positions of these sources, i.e., to their coordinates in a system defined by these sources alone, without reference to the ecliptic or the dynamical equinox. The important tasks of determining the orbits of the Earth, Moon, and planets, and the rotation of the Earth with respect to this extragalactic reference system to comparable levels of uncertainty will require more time. For these tasks, VLBI will also be important. Differential interferometric observations of extragalactic radio sources and spacecraft on the surfaces of, or in orbit around, the Moon and planets may be used to determine Earth-Moon and Earth-planet directions relative to the extragalactic sources (Counselman *et al.*, 1972; Counselman *et al.*, 1973).

* This pair was also observed on several other days, when the Sun was near 3C 279, in order to measure the solar gravitational deflection.

References

Clark, B. G.: 1973, *Proc. IEEE* **61**, 1242–1248.
Clark, T. A., Counselman, C. C., III, Hinteregger, H. F., Knight, C. A., Robertson, D. S., Rogers, A. E. E.,
  Shapiro, I. I., and Whitney, A. R.: 1973, *Bull. Am. Astron. Soc.* **5**, 30.
Cohen, M. H.: 1973, *Proc. IEEE* **61**, 1192–1197.
Counselman, C. C., III, Hinteregger, H. F., and Shapiro, I. I.: 1972, *Science* **178**, 607–608.
Counselman, C. C., III, Hinteregger, H. F., King, R. W., and Shapiro, I. I.: 1973, *Science* **181**, 772–774.
Knight, C. A.: 1973, personal communication.
Moran, J. M.: 1973, *Proc. IEEE* **61**, 1236–1242.
Rogers, A. E. E., Counselman, C. C., III, Hinteregger, H. F., Knight, C. A., Robertson, D. S., Shapiro, I. I.,
  and Whitney, A. R.: 1973, *Astrophys. J.* **186**, in press.
Schaper, L. W., Jr., Staelin, D. H., and Waters, J. W.: 1970, *Proc. IEEE* **58**, 272–273.

DISCUSSION

Elsmore: Am I right in thinking that atmospheric effects provide the limit to the positional accuracies that VLBI techniques can achieve?

Counselman: I think so, assuming that instrumental phase calibrators, which have been installed only recently, prove successful in reducing instrumental path-length uncertainty to less than a centimeter. If water-vapor radiometry techniques enable us, as we hope and expect, to determine atmospheric path length differences within 2 cm, then we should be able to determine source positions within milliseconds of arc using a 4000-km interferometer baseline. Of course, at such a level of angular resolution, source structure becomes important and must be specified in order to specify a source position.

Mulholland: Your determination of the variation of UT1 consistent with USNO to 5 ms should be very satisfying, since that is near the uncertainty level in the PZT determinations.

Counselman: Yes, the agreement between our determinations and those of the USNO is close enough to be satisfying, but not so close that you would be suspicious. Identical comments apply to our determinations of polar motion in relation to those by the BIH.

POSITION SOLUTION OF COMPACT RADIO SOURCES
USING LONG COHERENCE VLBI

J. S. GUBBAY, A. J. LEGG, and D. S. ROBERTSON

*Australian Defence Scientific Service, Dept. of Supply, Weapons Research Establishment,
Salisbury, South Australia*

Abstract. Compact components of several radio sources were observed in the course of VLBI observations at *S*-band between NASA-JPL Deep Space Stations located in Australia, South Africa and California, U.S.A., during the southern summer of 1971–72. These stations were equipped with H-maser frequency standards over this period so that the fringe frequency could be determined to better than 1 mHz.

The internal consistency of the position solution could be assessed for sources observed at least three times. This varied with source declination from $\pm 0''.1$ at a declination of $45°$ to $\pm 2''.0$ near the equator.

The spread in the position solution for P0727-11, a rapid variable at the observing frequency, was also large.

Positions obtained for some other sources have been used to assist in the identification of associated optical objects.

During the southern summer of 1971–72, two stations of the NASA-JPL Deep Space Network, DSS 51 at Hartebeestehoek, South Africa and DSS 41 at Island Lagoon, South Australia, co-operated in a series of three interferometer experiments (see Table I). G. Nicolson and P. Harvey of the South African CSIR were responsible for the conduct of the experiments at the South African terminal. This series was followed

TABLE I

Interferometer experiments using independent H-maser frequency standards

Epoch	Western Station designation	Eastern station designation
357–8 1971 21 1972 44–5 1972	DSS 51, Hartebeestehoek near Johannesburg, South Africa	DSS 41, Island Lagoon near Woomera, South Australia
52–3 1972	DSS 41	DSS 12, Goldstone, California, U.S.A.

within a few days by an experiment between DSS 12 near Goldstone in California and the Australian station, DSS 41. The three stations taking part in experiments across the two intercontinental baselines were equipped with 26 m antennae, maser front ends and hydrogen maser frequency standards.

Cohen and Shaffer took the results of a survey for compact components using a trans-Pacific baseline between the 64 m telescope at Goldstone, California and the 26 m telescope at Tidbinbilla, Australia, to determine the position of compact components. Their position determinations were among the first using VLBI techniques with independent local oscillators. The accuracy of these positions was limited by the stability of the station rubidium frequency standards and by the single parameter atmospheric model adopted.

Gliese, Murray, and Tucker, 'New Problems in Astrometry', 125–129. All Rights Reserved.

The experiments described here were part of a continuing study of the secular behaviour of the 13.1 cm emission from the compact components of radio sources. They were not designed to yield accurate source positions and station locations. Data was sampled at 48 kbits s^{-1} at a noise equivalent bandwidth of 14.3 kHz. The accuracy of group delay measurement is therefore limited to about 2 microseconds whereas the high stability (1 part in 10^{14}) of the frequency standard allows an accuracy of about 20 μHz in the phase-delay rate between the signals at the two stations. Consecutive dual observations, each of 660 s duration, were made on each source when its elevations from the respective stations were near optimum. However, three runs were made on each of three sources, including 3C 273 and 3C 279 in the first of the series of Southern Hemisphere, inter-continental baseline experiments. The corresponding position error triangles were consistent with a total reading error in the observed Doppler difference frequency of 0.5 mHz. Uncertainties in the atmospheric parameters, variations in the length of day and noise contribute to the total error. In addition, the orbital acceleration of the Earth between the respective arrival times at the two stations has not been taken into account.

These three sources were observed again during the second southern baseline viewing period. The observations of two of these sources were inter-leaved. Their position solutions however did not correspond with that obtained previously but it was found that the deviation in millihertz from the previous solutions followed a smooth curve suggesting a variable relative drift rate over the second half of the experiment (see Figure 1). Consequently, these results were ignored. The stability requirement of the S-band synthesizer for accurate position solutions is near its design limits. The synthesizer was found by Fanselow et al. (1971) to be a likely cause of variable drift. None of the sources observed in the first of this series was viewed in the third experiment.

The trans-Pacific experiment was conducted in conjunction with A.T. Moffet and D. Shaffer of the California Institute of Technology and D. Spitzmesser of the Jet Propulsion Laboratories. The only significant difference in equipment configuration from that adopted for the southern baseline experiments was the substitution of the S-band synthesizer by a multiplier chain. Three observations were taken on 3C 273 and four on 3C 279, the last of which was recorded when the source had dropped to an elevation of 5.8 deg from DSS 12 in California. Corrections for the atmosphere are based on a model ionosphere using values of total electron content obtained by F. Hibberd of the University of New England in Northern New South Wales and from the California Institute of Technology using geostationary satellites, and a model troposphere corrected for station altitude. As no ionospheric data was available from South Africa the variation in the ionosphere was inferred from the Australian results by introducing a delay equal to the difference in local time. Bottomside data now available indicate a correction to be applied to the delay. The atmospheric model was developed by K. Lynn of the Weapons Research Establishment. The position solutions for both sources appear to be internally consistent (see Figure 2). For the southern baseline and the trans-Pacific baseline, the separation in right ascension

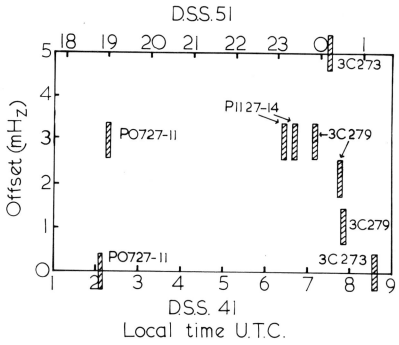

Fig. 1.   Relative bias for two trans-Indian Ocean experiments (day 347 1971 – day 21 1972).

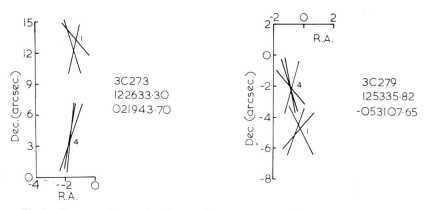

Fig. 2.   Source positions using H-maser frequency standard (R.A. in units of s × 15).
1 – South Africa-Australia 358 1971; 4 – Australia-U.S.A. 53 1972.

between these two sources are in fair agreement and are consistent with the separation
observed by Rogers *et al.* using VLBI across the United States at a wavelength of
3.8 cm. Agreement with the R.A. of the optical object associated with 3C 273 given
by Barbieri *et al.* (1972) and the R.A. of the optical object associated with 3C 279 is
within 0″.8 for either baseline. For each source, the difference in declination between

the two baseline solutions would result from the same relative error in length of baseline or the same difference in relative drift frequency. An adjustment of about -10 m to the trans-Pacific baseline and about -20 m to the southern baseline would yield good agreement with the source declinations determined by Rogers *et al.* (1974). However, an error of this magnitude would tend to be reflected in a right ascension error larger than the errors observed for the trans-Pacific baseline, if the station spin distances are correct. An alternative and more likely cause is a relative drift error of -5 mHz for the trans-Pacific baseline and -10 mHz for the southern baseline. These figures are consistent with a long term stability figure of 2 parts in 10^{-12} for the S-band synthesizer. This seems rather high but it should be noted that the relative drift rate appears to have halved for the trans-Pacific experiment where one station used a frequency multiplier instead of the synthesizer.

The Space Research Group at W.R.E. provides source positions to assist optical identifications at Mt Stromlo. Positions of objects beyond the viewing limits of northern interferometers are of particular importance but position determination of a low declination source has also been attempted. In this way, the optical object associated with OV 236, an optical and radio variable, was identified by B. Peterson of Mt Stromlo observatory. With the commissioning of the 64 m antenna at Tidbinbilla, Australia, a larger number of sources of possible optical interest may be observed.

Three observations were made on 3C 279 by DSS 43, the 64 m antenna at Tidbinbilla, and DSS 11 in California. The use of Rubidium frequency standards limited the period of phase coherent integration to 60 s with a consequent increase in random error. However, although different stations were used in this later experiment, values of R.A. of 3C 279 for the two trans-Pacific experiments about 13 months apart are in good agreement. The results from the two trans-Pacific baseline experiments suggest that the true baseline is about $0''.8$ eastwards of the assumed direction, if the compact component coincides with the optical object. The origin of the co-ordinate system for 3C 279 in Figure 2 is at the position for the optical object given by Kristian and Sandage (1970). The difference between the true and assumed directions of the southern baseline appears to be smaller but in the same sense. The sense of the required corrections to the two baselines is opposite to that obtained by Cohen and Shaffer (1971) for the Tidbinbilla-Goldstone baseline involving DSS 42, a 26 m telescope, less than 300 m north of DSS 43, the 64 m telescope at Tidbinbilla and DSS 14 at Goldstone. The difference between the location solution LS 25 and that adopted in this work LS 35 would account for a difference in the correction to the hour angle of the baseline of less than $0''.1$. The authors concluded however that the correction obtained then could not be used to improve the station locations.

The equatorial projections of the two baselines, California-Australia-South Africa, span $216°$ of longitude. The maximum error in the Hour Angle of the California (DSS 12 or DSS 11) – South Africa (DSS 51) baseline can be deduced from the equatorial projections and errors in hour angle of the other two baselines forming a triangle in the equatorial plane. Thus the hour angle of a source can be determined to better than 1″.0, even when only one source is observed, using any of these NASA-JPL baselines. The results of all the sources observed in this series will be used to determine the errors in the baselines, the source declinations and the rate offsets for each experiment.

References

Barbieri, C., Capaccioli, M., Ganz, R., and Pinto, G.: 1972, *Astron. J.* **77**, 444.

Cohen, M. H. and Shaffer, D. B.: 1971, *Astron. J.* **76**, 91.

Fanselow, J. L., MacDoran, P. F., Thomas, J. B., Williams, J. G., Finnie, C. J., Sato, T., Skjerve, L., and Spitzmesser, D. J.: 1971, JPL Tech. Rept. 32–1526. Vol. V, pp. 45–57.

Kristian, J. and Sandage, A.: 1970, *Astrophys. J.* **162**, 391.

Rogers, A. E. E., Counselman, C. C., III, Hinteregger, H. F., Knight, C. A., Robertson, D. S., Shapiro, I. I., Shitney, A. R., and Clark, T. A.: 1974, submitted to *Astrophys. J. Letters*.

ACCURACY ESTIMATION OF ASTRONOMICAL CONSTANTS FROM LONG BASELINE INTERFEROMETER OBSERVATIONS*

H. G. WALTER

European Space Operations Centre of ESRO, Darmstadt, F.R. Germany

Abstract. Inherent in Very Long Baseline Interferometry (VLBI) is the potentiality of determining relative angular positions of radio sources to an accuracy which is superior to present methods by at least one order of magnitude. For VLBI observations such as time delay, fringe frequency and fringe phase the sensitivity to the uncertainties of astronomical constants and geophysical parameters, i.e. luni-solar precession, nutation constant, rate of change of obliquity, polar motion and Earth rotation, is investigated.

The partial derivatives of the observations with respect to these unknowns form the core of the estimation process and give evidence of the influence of the source position and the baseline geometry on their magnitudes. By means of a variance-covariance analysis the standard deviations and the cross correlation coefficients of the unknowns are derived. It is inferred from them that VLBI observations are not only sensitive to variations in the astronomical constants and geophysical parameters, but that the effects caused by these variations are also measurable and separable, if the observations are extended over a sufficiently long interval of time, and the baselines and source positions are chosen appropriately.

DISCUSSION

Fricke: Can you say how the accuracy of the determinations changes, if the number of unknowns is decreased, in particular, if you don't include as an unknown the rate of change of the obliquity?

Walter: Since the unknowns are usually correlated, an apparent improvement of accuracy is obtained if a smaller number of unknowns is processed. Moreover, the large time spacing of observations over 10 years or more would be avoidable. In order to arrive at a meaningful error estimation, however, the number of unknowns should not be unnecessarily diminished.

---

* The full text of this paper is being published as *Bull. GRGS*, No. 10 of the Groupe Recherche Géodesié Spatiale, Observatoire de Paris, Meudon, France.

RADIO AND OPTICAL ASTROMETRY

(Invited Paper)

C. M. WADE

National Radio Astronomy Observatory*, Green Bank, W.Va., U.S.A.

Abstract. Radio positional measurements have achieved an accuracy as high as that of optical astrometry, with uncertainties no greater than a few hundredths of an arc second in each coordinate. Declinations and relative right ascensions are determined absolutely. Since the principal sources of systematic error are different for radio and optical astrometry, radio measurements can be useful in the preparation of future fundamental catalogues.

1. Introduction

A quarter of a century ago, the most accurate measurements of radio source positions were uncertain by several minutes of arc. The quality of radio work has since improved to the point that it now is possible to make systematic positional measurements of small-diameter sources which are as accurate as the fundamental optical catalogues. Interferometric methods developed at the Mullard Radio Astronomy Observatory in Great Britain and at the National Radio Astronomy Observatory in the United States yield precise absolute declinations and right ascension differences. It still is necessary to use optical information to fix the origin of the right ascensions, since no object in the solar system is suitable for radio astrometry. In all other respects, the radio measurements are fundamental in the astrometric sense.

Radio astrometry has several important advantages in comparison with optical astrometry. The effects of atmospheric refraction are relatively unimportant. Large angles can be measured with about the same accuracy as small ones, so regional systematic errors are not a serious problem. Observations of a large number of sources well distributed on the sky can be analyzed simultaneously in a way that fixes most of the instrumental parameters along with the source positions, so the measurements are largely self-calibrating. The observations determine the radii of the diurnal circles of the sources, thus referring the declinations directly to the Earth's instantaneous axis of rotation. Most sources of small angular size are extragalactic, and a grid of fundamental positions which are not affected by proper motions can be established readily. Finally, the observations can be made rapidly, and analyzed entirely by electronic computers.

We shall consider here only the most recent fundamental radio interferometric measurements. Much excellent radio work has been done by reference to optically determined calibration positions, but such measurements are outside the scope of the present paper since they are essentially differential. We shall also ignore lunar occulta-

---

* Operated by Associated Universities, Inc., under contract with the National Science Foundation.

Gliese, Murray, and Tucker, 'New Problems in Astrometry', 133–139. All Rights Reserved.
Copyright © 1974 by the IAU.

tion measurements because their accuracy is limited by instrumental noise and the uncertainty of the lunar limb corrections.

2. Fundamental Radio Measurements

Radio interferometric techniques for absolute position measurements have been developed independently at the Mullard Radio Astronomy Observatory (Elsmore and Mackay, 1969; Smith, 1971; Ryle and Elsmore, 1973) and at the National Radio Astronomy Observatory (Wade, 1970; Brosche *et al.*, 1973). Although the underlying principles of the measurements at the two observatories are much the same, the actual methods differ considerably because the instruments are quite dissimilar (Ryle, 1972; Hogg *et al.*, 1969). The difference which has the greatest influence on method is in the baseline orientations. Some of the characteristics of the interferometers are summarized in Table I.

TABLE I

Characteristics of the Mullard and NRAO interferometers

	Mullard	NRAO
Number of antennas	8	3
Antenna diameter	13 m	26 m
Maximum baseline	4.6 km	2.7 km
Baseline azimuth	93°19′	62°02′
Latitude	52°.2 N	38°.4 N
Operating wavelength	6 cm	4+11 cm
Min. fringe spacing	2″.7	2″.9

The accuracy of both instruments is limited primarily by random fluctuations in the signal paths through the troposphere (and to a lesser extent, through the ionosphere). The effect is analogous to optical scintillation, although the characteristic time scale is longer by two to three orders of magnitude. Since the process is random, the errors can be reduced considerably by averaging the results of repeated observations. The uncertainties due to instrumental noise are negligible except for very weak sources.

Atmospheric refraction is not an important source of error in radio astrometry, since the positional information lies in the relative arrival times of the signals at the different antennas rather than in the apparent direction of arrival. Small corrections are necessary to compensate for the slightly differing elevations of the antennas above sea level and for the curvature of the atmosphere, but these are easily calculated and introduce little uncertainty.

The Mullard and NRAO methods both determine declinations absolutely, since in effect they measure the radii of the diurnal circles of the sources. Differences of right ascension between sources are also found absolutely, with an accuracy which is independent of the magnitude of the difference. Unfortunately, the angular sizes of the Sun and the planets are too large for their positions to be measured accurately with radio interferometers, and there is therefore no way to refer the right ascensions directly to the vernal equinox by purely radio methods. Instead, it is necessary that the observations include at least one object whose right ascension has been found by optical means. The Mullard right ascensions are referred to the FK4 position of β Persei, which is an intermittent radio source (Wade and Hjellming, 1972). At NRAO, the right ascensions have been adjusted to agree in the mean with optical measurements of a large number of sources, mostly quasars (Kristian and Sandage, 1970; Murray et al., 1971).

The major source of systematic error is imperfect instrumental calibration, particularly in the determination of the collimation error (which reflects departures from symmetry in the paths of the signals collected by the different antennas). Small changes in baseline length and orientation due to differing thermal effects at the various antennas can also cause diurnal errors which are difficult to evaluate. Careful calibration procedures can keep such errors small, but they cannot be eliminated entirely.

Accuracies of a few milliseconds in right ascension and a few hundredths of an arc second in declination have been achieved at Mullard (Ryle and Elsmore, 1973). This is fully as good as the FK4. Accuracies of about 0″.15 have been attained at NRAO (Brosche et al., 1973). The higher precision of the Mullard measurements is the result of averaging a large number of observations, which much reduces the phase noise caused by atmospheric inhomogeneities. The instruments and methods used at the two observatories are inherently capable of similar accuracy.

3. Comparison of Radio and Optical Results

Several fairly extensive lists of accurate optical positions of radio sources have been published in the last 5 years (e.g., Bolton, 1968; Kristian and Sandage, 1970; Hunstead, 1971; Véron, 1972; Barbieri et al., 1972). Perhaps the most accurate measurements have been made at Cambridge (Argue and Kenworthy, 1972; Argue et al., 1973). These are essentially on the AGK3 system, although they are based partly on earlier measurements in the FK4 system by Murray et al. (1969, 1971).

We shall restrict the present discussion to comparing the Cambridge optical positions with the most recent fundamental radio measurements at Mullard and NRAO (Ryle and Elsmore, 1973; Brosche et al., 1973). Table II gives the measured positions for 1950.0 with their mean errors; the errors of the optical positions are the external errors. The NRAO right ascensions have been adjusted slightly to agree in the weighted mean with the optical values.

It is clear from Table II that the agreement between the three sets of measurements is generally good. This is confirmed by Table III, which gives the weighted (by the inverse squares of the mean errors) average differences between the sets. The overall systematic agreement is at least as good as 0″.1, which is less than the claimed mean errors of the individual NRAO and optical measurements.

TABLE II

Radio and optical positions (1950.0) for 14 radio sources

Source	Radio Mullard	Radio NRAO	Optical
0056-00		$31^{s}761 \pm 0^{s}020$	$31^{s}771 \pm 0^{s}017$
$-00°09'$		$18''72 \ \pm 0''52$	$18''08 \ \pm 0''30$
3C48	$49^{s}827 \pm 0^{s}002$	$49^{s}817 \pm 0^{s}012$	$49^{s}819 \pm 0^{s}012$
01^h34^m $+32°54'$	$20''63 \ \pm 0''04$	$20''74 \ \pm 0''18$	$20''40 \ \pm 0''12$
3C84	$29^{s}562 \pm 0^{s}003$		$29^{s}548 \pm 0^{s}014$
03^h16^m $+41°19'$	$51''99 \ \pm 0''04$		$52''19 \ \pm 0''12$
3C138	$16^{s}526 \pm 0^{s}005$	$16^{s}520 \pm 0^{s}011$	$16^{s}521 \pm 0^{s}011$
05^h18^m $+16°35'$	$27''06 \ \pm 0''09$	$27''29 \ \pm 0''25$	$27''06 \ \pm 0''15$
3C147	$43^{s}503 \pm 0^{s}003$	$43^{s}498 \pm 0^{s}016$	$43^{s}492 \pm 0^{s}016$
05^h38^m $+49°49'$	$42''87 \ \pm 0''02$	$42''94 \ \pm 0''15$	$43''10 \ \pm 0''12$
0736+01		$42^{s}510 \pm 0^{s}017$	$42^{s}496 \pm 0^{s}010$
07^h36^m $+01°44'$		$00''29 \ \pm 0''44$	$00''29 \ \pm 0''21$
4C39.25		$55^{s}296 \pm 0^{s}014$	$55^{s}310 \pm 0^{s}015$
09^h23^m $+39°15'$		$24''30 \ \pm 0''17$	$22''94 \ \pm 0''14$
3C279		$35^{s}821 \pm 0^{s}021$	$35^{s}842 \pm 0^{s}018$
12^h53^m $-05°31'$		$07''36 \ \pm 0''30$	$07''28 \ \pm 0''30$
3C286	$49^{s}653 \pm 0^{s}004$	$49^{s}681 \pm 0^{s}012$	$49^{s}657 \pm 0^{s}012$
13^h28^m $+30°45'$	$58''79 \ \pm 0''05$	$58''32 \ \pm 0.19$	$58''46 \ \pm 0''12$
3C309.1	$56^{s}644 \pm 0^{s}005$	$56^{s}616 \pm 0^{s}038$	$56^{s}648 \pm 0^{s}033$
14^h58^m $+71°52'$	$11''17 \ \pm 0''02$	$11''26 \ \pm 0''16$	$11''33 \ \pm 0''16$
3C345	$17^{s}603 \pm 0^{s}002$	$17^{s}609 \pm 0^{s}014$	$17^{s}606 \pm 0^{s}013$
16^h41^m $+39°54'$	$10''89 \ \pm 0''03$	$11''26 \ \pm 0''16$	$10''72 \ \pm 0''12$
BL Lac	$39^{s}362 \pm 0^{s}007$	$39^{s}370 \pm 0^{s}014$	$39^{s}397 \pm 0^{s}021$
22^h00^m $+42°02'$	$08''69 \ \pm 0''04$	$08''77 \ \pm 0''16$	$08''47 \ \pm 0''14$
CTA 102	$07^{s}793 \pm 0^{s}005$		$07^{s}823 \pm 0^{s}029$
22^h30^m $+11°28'$	$22''89 \ \pm 0''20$		$23''07 \ \pm 0''25$
3C454.3	$29^{s}510 \pm 0^{s}006$		$29^{s}533 \pm 0^{s}011$
22^h51^m $+15°52$	$54''54 \ \pm 0''09$		$54''98 \ \pm 0''15$

TABLE III

Weighted mean differences

	Mullard-NRAO	Mullard-Optical	NRAO-Optical	Notes
$\langle \Delta\alpha \rangle$	$-0^{s}002 \pm 0^{s}006$	$-0^{s}002 \pm 0^{s}005$	–	(1)
$\langle \Delta\alpha \cos\delta \rangle$	$-0''01 \ \pm 0''07$	$-0''01 \ \pm 0''05$	$-0''01 \pm 0''05$	
$\langle \Delta\delta \rangle$	$-0''08 \ \pm 0''09$	$-0''01 \ \pm 0''08$	$+0''13 \pm 0''09$	(2)
No. of sources	7	10	10	

Notes:

(1) The NRAO right ascensions were adjusted to make the weighted mean difference from the optical equal to zero.

(2) 4C39.25 was omitted because the NRAO and optical declinations differ by $1''36$, which is excessive in relation to the differences for the other sources.

The number of sources in Table II is far too small to permit a search for possible regional differences between the radio and optical positions. Even in this small sample, however, there appears to be a consistent trend in the difference between the Mullard and optical right ascensions, as a function of right ascension. This can be seen in Figure 1. The right ascensions of the sources common to the Mullard and optical lists

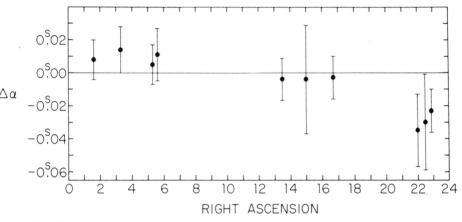

Fig. 1.   Distribution of the differences between the Mullard and optical right ascension measurements, as a function of right ascension.

fall into three loose groups. Within each group, the values of $\Delta\alpha$ are about the same, but there is a noticeable difference from group to group. The effect appears to be real, but one cannot say at this time whether the source of the discrepancy is in the optical or the radio measurements. The NRAO measurements are not accurate enough to resolve the matter with any certainty.

4. Conclusions

Radio astrometry, although new, has achieved a level of accuracy which rivals the best optical work. It has several valuable advantages over existing optical astrometric methods, of which the most important are:

(a) the relative unimportance of atmospheric refraction;

(b) the ability to measure absolute declinations which are automatically referred to the Earth's instantaneous axis of rotation; and

(c) the ability to measure large angles with essentially the same accuracy as small angles.

The principal disadvantage, from the standpoint of fundamental astrometry, is the inability of existing instruments to determine the location of the vernal equinox without recourse to optical data. This is a temporary deficiency, however, since the Very Large Array (VLA) now under construction in the United States will be sensitive enough to observe the brightest minor planets.

The major source of random error in radio astrometry is the phase instability caused by tropospheric inhomogeneities. Under average conditions, the accuracy attainable in a single observation with a single pair of antennas is limited to about 0."2. The error can be reduced considerably by taking the mean of a large number of observations. This is shown by the much higher accuracy of the Mullard measurements in Table II as compared to the NRAO measurements. The Mullard positions are the result of many observations with 16 antenna pairs, while the NRAO positions are derived from single observations (each consisting of 5 min of integration at three widely separated hour angles) with 3 antenna pairs.

The systematic errors of radio astrometry are due to imperfect instrumental calibration. In a well-designed interferometer system, all of the relevant parameters except the collimation error are highly stable, and most are readily calibratable from radio observations alone. The collimation error is troublesome, however, since it is the result of unstable asymmetries in the electronic system. Its effects can be minimized by careful design of the observing program.

Most of the radio sources which are suitable for astrometric measurement with interferometers are extragalactic, mostly quasars and the nuclei of active galaxies. These objects are numerous and well distributed over the sky. Thus radio astrometry can readily establish a set of precise extragalactic reference points for the differential measurement of proper motions. The faintness of the optical counterparts of most of these objects (15th to 19th mag.) will cause some difficulty in their practical employment as proper motion standards, but the problem is not fundamental. Finally, repeated measurement of the positions of extragalactic sources over a sufficiently long span of time will permit a refinement of the precessional constants without the complications caused by proper motions and galactic rotation.

Radio observations can make valuable contributions to the field of astrometry as a whole, particularly in establishing a uniform positional reference frame over the entire sky. I believe that radio astrometry will soon play an integral part in the construction of fundamental catalogues, because of its ability to give results which are free of regional systematic errors as well as its inherently high precision. It is unfortunate that there is no radio interferometer in the southern hemisphere at present which is usable for astrometric work. Such an instrument would be invaluable in tying the astrometric systems of the two hemispheres together.

References

Argue, A. N. and Kenworthy, C. M.: 1972, *Monthly Notices Roy. Astron. Soc.* **160**, 197.
Argue, A. N., Kenworthy, C. M., and Stewart, P. M.: 1973, *Astrophys. Letters* **14**, 99.
Barbieri, C., Capaccioli, M., Ganz, R., and Pinto, G.: 1972, *Astron. J.* **77**, 444.
Bolton, J. G.: 1968, *Publ. Astron. Soc. Pacific* **80**, 5.
Brosche, P., Wade, C. M., and Hjellming, R. M.: 1973, *Astrophys. J.* **183**, 805.
Elsmore, B. and Mackay, D. C.: 1969, *Monthly Notices Roy. Astron. Soc.* **146**, 361.
Hogg, D. E., Macdonald, G. H., Conway, R. G., and Wade, C. M.: 1969, *Astron. J.* **74**, 1206.
Hunstead, R. W.: 1971, *Monthly Notices Roy. Astron. Soc.* **152**, 177.
Kristian, J. and Sandage, A.: 1970, *Astrophys. J.* **162**, 391.

Murray, C. A., Tucker, R. H., and Clements, E. D.: 1969, *Nature* **221**, 1229.
Murray, C. A., Tucker, R. H., and Clements, E. D.: 1971, *Roy. Obs. Bull.*, No. 162.
Ryle, M.: 1972, *Nature* **239**, 435.
Ryle, M. and Elsmore, B.: 1973, *Monthly Notices Roy. Astron. Soc.*, in press.
Smith, J. W.: 1971, *Nature Phys. Sci.* **232**, 150.
Véron, M. P.: 1972, *Astron. Astrophys.* **20**, 471.
Wade, C. M.: 1970, *Astrophys. J.* **162**, 381.
Wade, C. M. and Hjellming, R. M.: 1972, *Nature* **235**, 270.

DISCUSSION

Eichhorn: I am not particularly worried about not all of the error boxes overlapping since the true values of a quantity will be within one sigma in only about two of three cases.

Wade: In fact the rate of agreement within the claimed mean errors is about two thirds, as one would expect.

Eichhorn: I am also happy about the apparent approaching demise of the vernal equinox as a fundamental point in astrometry. The vernal equinox is, after all, defined only by the kinematics – and only very incidentally the dynamics – of the Earth rotation and revolution. It is thus very intimately tied to the Earth and the solar system, and really not at all germane to matters galactic and cosmological.

Van Herk: How well should optical positions coincide with radio sources? It has been said that the position of the radio sources depends on the frequency. Have radio sources to be 'compact' to have the two positions (optical and radio) coinciding?

Wade: It is hard to give a general answer, since the radio sources are highly diverse in their properties. In cases where there is a spectral gradient across the source, the radio position will of course be a function of wavelength. When the spectrum is quite uniform over the source, such effects should be negligible. This is true also when the angular extent of the source is appreciably smaller than the error of the positional measurements.

Baars: Yesterday Prof. Bok urged radio astronomers to find all radio stars. I would like to inform you of a project in which Dutch and German radio astronomers use the 100-m telescope in Bonn to search for radio radiation at 3 cm wavelength. After a possible detection an accurate position is determined with the Synthesis Radio Telescope in Westerbork at 6 cm wavelength. Several stars have been detected in Bonn, while the Westerbork follow-up is now being made.

I have one new radio star to report, detected by Dr H. Wendker (Hamburg Observatory) and myself with the Westerbork telescope. It is P Cygni and it shows a flux density at 6 cm wavelength of 10^{-28} Wm^{-2} Hz^{-1}. The derived radio position lies within $0''.6$ from the AGK3 position of the star. The estimated error in the radio position is $\pm 0''.5$.

Brosche: As long as radio astrometry is using the present constant of precession, 'artificial' proper motions have to be introduced also for extragalactic objects due to the necessary correction of this constant.

THE TEXAS RADIO ASTROMETRIC SURVEY

JAMES N. DOUGLAS

University of Texas, Austin, Tex., U.S.A.

Abstract. The 5-element Texas Interferometer is engaged in a 365 MHz survey of the sky with a primary goal of establishing positions of about 50 000 discrete radio sources with an accuracy of about 1″ in each coordinate. Measurements are made relative to optical positions of identified sources, and the 2000 positions thus far obtained support our expectations of the accuracy of the survey. In a companion program, optical positions of objects near radio source positions are being measured to $\pm\frac{1}{2}''$ accuracy on glass copies of the Palomar Sky Survey, yielding both improved overall calibration of the radio positions and identification of associated optical objects on the basis of position coincidence alone, without the selection effects usually introduced by auxiliary identification criteria.

Gliese, Murray, and Tucker, 'New Problems in Astrometry', 141. All Rights Reserved.
Copyright © 1974 by the IAU.

ABSOLUTE POSITIONS OF RADIO SOURCES OBTAINED WITH THE NRAO INTERFEROMETER*

P. BROSCHE

Astronomische Institute, Bonn, F.R. Germany

Abstract. Measurements by C. M. Wade and R. M. Hjellming with the three-element-interferometer at Green Bank were reduced simultaneously for instrumental parameters and positions of 59 radio sources. The minor axes of the confidence ellipses have typical values of about $0\rlap{.}''15$ while the major axes vary with declination from $0\rlap{.}''15$ to $1\rlap{.}''4$.

* Published in *Astrophys. J.* **183**, 805 (1973).

DETERMINATION OF AN EXTRAGALACTIC RADIO SOURCE POSITION CATALOGUE SUITABLE FOR USE IN MONITORING GEOPHYSICAL PHENOMENA

JOHN L. FANSELOW and J. G. WILLIAMS

Jet Propulsion Laboratory, California Institute of Technology, Pasadena, Calif., U.S.A.

Abstract. The use of radio interferometric techniques to monitor geophysical pheno-mena such as UT1-UTC variations, polar motion, and tectonic plate motion requires that there be available a fairly uniform distribution of extragalactic radio sources whose relative positions are known to better than 0″.01. These sources must also have very small proper motion, and not be subject to asymmetric changes in radio structure or to very large variations in intensity. In addition, to facilitate tying this frame of reference to the optical star frames, it is desirable that many of these radio objects have optical counterparts. This paper describes our progress in obtaining such a frame and sets forth our desiderata for further work by radio and optical astronomers.

DISCUSSION

Murray: As an optical astronomer in the radio source position business I would like to know when we can expect a definitive list of compact sources, at say 0″.02, which are suitable for position calibration. Otherwise we waste large-telescope time getting plates on useless sources.

Williams: About 75 sources have been seen at our wavelength (13 cm) in very long baseline interferometry experiments. These are sources which have at least one compact component at the $\leqslant$0″.002 level. They may, however, have larger scale structure too. We will make a list of these sources available to interested workers. Of these sources 25 have been used to make a preliminary radio catalogue based on the interfero-metry data.

Walter: What is the distribution in declination of the catalogued radio sources?

Williams: From $-16°$ to $+69°$ in the extreme. Most of the sources lie between $0°$ and $+50°$.

DISCUSSION AFTER SESSION C

Douglas: With reference to selecting radio source counterparts for optical astrometric measurement, angular structure enters in two ways:

(1) If angular structure is large compared to the scale of interferometer fringes, the radio centroid is not measurable; such sources should not be chosen.

(2) Finite angular structure which is small compared to interferometer fringes will not impede measurement of radio brightness centroid. However, it may alter the confidence with which one assumes radio and optical centroids to be coincident. In fact, for an individual source, such an assumption is never fully safe, so radio-optical comparisons should be statistical.

Elsmore: I agree with Dr Douglas but I would add that sometimes the optical object is extended: in particular, 3C 48 has a jet to the north of this object, as shown in plates by Sandage. It is interesting to note that Dr Wade showed that both the NRAO and Cambridge radio positions lie north of the optical centroid.

Moffet: As a partial answer to Mr Murray's question (as to which radio sources have compact components), we have found that essentially all sources with pronounced centimeter-excess radio spectra have components of the order of $0\rlap{.}''001$ diam. In the absence of convincing evidence to the contrary, we assume that the compact radio and optical components of such objects are coincident. At longer wavelengths the centroids of the radio sources may be displaced because asymmetric large-scale components are also present, as is the case for 3C279. Mr Elsmore has already mentioned that extended components may similarly affect optical positions. Nevertheless it is important to have good optical positions for the compact components of even very complex objects such as the nuclei of Seyfert galaxies (for example NGC 1275 and 3C 120).

Tucker: Does intergalactic dispersion separate the positions of the radio source and the optical counterpart?

Wade: There is no evidence of such dispersive shifts to the $0\rlap{.}''1$ level, although it has been sought.

Gliese, Murray, and Tucker, 'New Problems in Astrometry', 147. All Rights Reserved.

ASTROMETRY WITH LARGE TELESCOPES

PROPER MOTION AND PARALLAX PROGRAMMES
FOR LARGE TELESCOPES

(Invited Paper)

C. A. MURRAY

Royal Greenwich Observatory, Herstmonceux, England

Abstract. Proper motion work at the present time relies on the availability of plates taken in the past, for which the magnitude limit is generally not fainter than about $m_{pg} = 16$, apart from Luyten's survey with the Palomar Schmidt. It is very important for future studies to fainter magnitudes in, for example, star clusters, that sufficient time is made available on large telescopes so that plates can be taken now. Consideration should also be given to an extension of proper motion work in Selected Areas.

Ideally, trigonometric parallax observations are best carried out on a dedicated telescope, such as the USNO 61-in. reflector, with a large amount of time available; but if sufficient observational material is to be obtained on faint stars, time must also be taken on other medium sized and large reflectors. With the severe limitation of time available on these telescopes, the criteria for selection of stars for such programmes become of prime importance. Large Schmidt telescopes, with fast automatic plate measurement, may possibly be used for surveys of nearby stars, unbiased by any kinematic or photometric selection effects.

1. Introduction

The large output of astrometric data during the first decades of this century, notably the photographic positional catalogues and parallax work on long focus refractors, has provided a reservoir of information for objects brighter than about the twelfth magnitude. Much of these data were collected without a specific astrophysical motivation, but they have nevertheless proved invaluable in subsequent researches. Nowadays however the astrometrist is at somewhat of a disadvantage; with the increase in the number of large telescopes now available, the interest in astrophysics is shifting towards fainter apparent magnitudes for which there are as yet little astrometric data.

Astrometrists must also use these larger telescopes to obtain their raw material, but, while the technique of obtaining good plates is relatively simple, they must have a sufficient lapse of time for measuring a parallax and even longer for precise proper motion. It is therefore important that good observational programmes be planned now and plates taken as soon as possible.

Astrometry itself has been responsible for the discovery of many interesting objects, primarily through proper motion surveys, but as these surveys go to fainter and fainter magnitudes, such as the current survey by Luyten with the 48-in. Schmidt, the problem of obtaining precise parallax and proper motion data becomes more acute.

In the present context I shall define 'large' to mean an aperture of about 48 in. or greater. Considerable experience has already been gained with astrometric work on telescopes in this category. Strand (1962) has discussed the parallax work by van Maanen with the 60-in. and 100-in. reflectors on Mount Wilson. At Herstmonceux we have also used 60-in. Cassegrain plates successfully for proper motions in some cluster fields (e.g. Murray *et al.*, 1965; Cannon and Lloyd, 1970) and I have also used

Gliese, Murray, and Tucker, 'New Problems in Astrometry', 151–158. All Rights Reserved.

old 60-in. Newtonian photometric plates on Selected Areas (Murray and Sanduleak, 1972) in spite of the comatic images. Furthermore, an investigation of the astrometric performance of the 98-in. Isaac Newton Telescope with the Wynne four element prime focus corrector, has shown that precise work is possible with this instrument (Murray, 1971). The 48-in. Palomar Schmidt has also been shown to be capable of good astrometry. In addition to the blink survey, Luyten (1963) has used plates from this telescope for measuring a parallax. He has also reported satisfactory positional measurements over the whole area of a 14 in. sq plate (Luyten and La Bonte, 1972) in a southern field using SAO catalogue positions for reference stars and, likewise at Herstmonceux a positional investigation has been carried out on a northern field over a 10 in. sq plate taken with the same telescope, using AGK3 stars as references (Parkes and Penston, 1973). In both these investigations the major contribution to the errors is from the catalogue positions, but the actual errors (0.″6 for the SAO field and 0.″36 for the AGK3 field) are reasonable, bearing in mind the likely random errors in the star places and the accidental errors of measurement.

My own feeling is that, given good plates taken with whatever optical system, the information is there, and it is up to the ingenuity of the astrometrist, aided by the computer and measuring machine to extract the information from the plates.

2. Trigonometric Parallaxes

A major advance in the astrometric study of stars fainter than about twelfth magnitude has been brought about through the operation of the USNO 61-in. astrometric reflector at Flagstaff. In a paper presented at the symposium associated with the dedication of that telescope, Worley (1966) summarized the state of parallax work at that time; in particular, parallaxes had then been measured for fewer than 100 stars in the northern sky fainter than $m_v = 13$. Since then the situation has changed dramatically; out of the two hundred stars for which parallaxes from 61-in. reflector observations have been published (Riddle, 1970; Routly, 1972), 75% are fainter than $m_v = 13$, but only four are fainter than $m_v = 16$. Also, van Altena (1971) has published a few results from the new Yerkes programme for stars of thirteenth and fourteenth magnitude.

At Herstmonceux we are making a small contribution to this problem with a programme of some 20 faint stars between photographic magnitudes 16 and 18.5, which we are observing at the prime focus of the Isaac Newton Telescope. However, this telescope is used for many other purposes and my group gets, on the average, an allocation of just over three full nights per month, not all of which are clear, for this and other astrometric programmes. In our case it is therefore very important to choose the programme carefully to avoid wasting time on small parallaxes. For the current programme the selection has been mostly on proper motion criteria, but I am not too sure, even at these faint magnitudes, that this was the wisest course.

It is very well known that, at least at brighter magnitudes, selection by proper motion introduces a bias in favour of high velocity stars with small parallaxes. This is

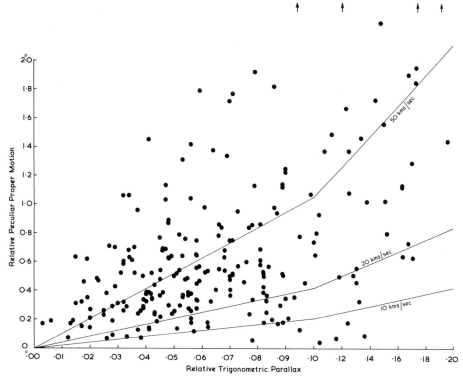

Fig. 1.   Plot of total peculiar proper motion (ordinate) against trigonometric parallax (abscissa) for stars in Vyssotsky's catalogues for which parallaxes are given.

well illustrated in Figures 1 and 2 which show plots of relative parallax versus peculiar proper motion (i.e. corrected for standard solar motion) for two samples of K–M dwarfs from Vyssotsky's objective prism survey. Figure 1 shows the data for all stars in the catalogues for which parallaxes had already been measured at the time of the survey, and hence had been mostly selected on grounds of large proper motion, whereas Figure 2 shows the data for stars measured subsequently at Van Vleck Observatory and at Herstmonceux for which parallaxes had not been measured prior to the spectral survey. In both figures the straight line loci represent tangential velocities of 10, 20, and 50 km s^{-1}. In spite of the small number of stars in Figure 2, it is immediately clear that the relative proportions of stars with different total transverse velocieties is markedly different in the two selections. Taking the parallax range 0″.04 to 0″.10 as representative, the actual numbers are:

Transverse velocity (km s^{-1})	Figure 1 (Previous parallaxes)	Figure 2 (New parallaxes)
≤ 10	4	8
10–20	17	11
20–50	69	12
> 50	50	1

Extropolation of small numbers is of course dangerous, but these figures are not in-consistent with the existence of a large number of K–M dwarfs with low velocities. That the bias is present among fainter stars also, is illustrated in Figure 3, which shows the same data for stars in the two catalogues of the 61-in. results for which $B-V>1.35$ which are presumably also M dwarfs and had generally been selected from Giclas' survey. Clearly some selection criterion independent of proper motion is essential in planning a parallax programme of faint stars on a telescope on which limited observing time is available. An obvious course would be to carry out a preliminary photometric survey in, say, R and I to find the red stars. Those with apparent magnitudes fainter than about $m_{pg}=12$ in high latitudes will most probably

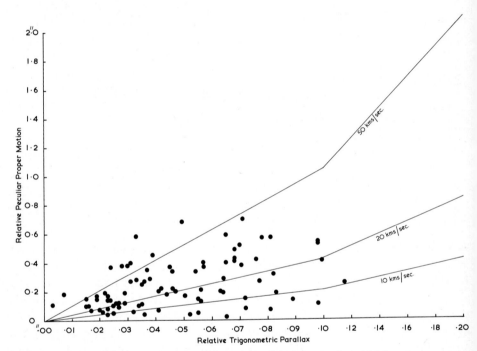

Fig. 2. As Figure 1, for stars without parallaxes in Vyssotsky's catalogue but which have been measured subsequently at van Vleck˙and Herstmonceux.

be dwarfs, but the luminosity class can be estimated by means of the four colour technique of D. H. P. Jones (1973). This may be a laborious process, but it is not as laborious as the accumulation of sufficient plates for a parallax determination.

Alternatively, objective prism techniques such as that used by Sanduleak (1964) in the north galactic pole region could be used to detect M stars to about $m_v=16.5$, or perhaps fainter on the large Schmidt telescopes.

White dwarfs and other sub-luminous stars are also obvious candidates for parallax work. Greenstein and Eggen (1966), in their contribution to the Flagstaff symposium, proposed a list of some 60 white dwarfs for which parallaxes were needed. Already about 25% of these have been observed with the 61-in. reflector. As Strand and Riddle (1970) have pointed out, several stars in the Giclas survey, with intermediate colours, which were put on the 61-in. programme as white dwarf suspects, have turned out to be high velocity stars with small parallaxes, and they urge that in future, such stars should also be first studied photometrically before parallax observations are started.

However, even after rigorous screening by photometric and spectroscopic observations, the lists of large proper motion stars by Giclas and, for the very faint stars, by Luyten, will eventually provide many suitable candidates for parallax observations.

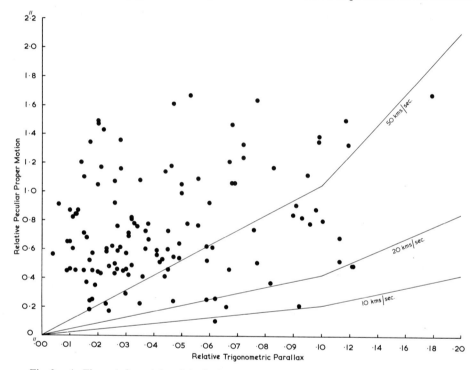

Fig. 3.   As Figure 1, for red dwarfs in the first two catalogues of results from the 61-in. reflector.

The search for nearby stars has hitherto invariably been bedevilled by selection effects. However, with modern fast plate measuring techniques becoming available, an entirely new attack on the problem is possible. The area which can be photographed on a 14 in. sq plate, with the 48 in. Palomar Schmidt and the UK Schmidt on Siding Spring is about a thousandth of the whole celestial sphere. Even at present, more than a thousand stars are known to be within 20 pc of the Sun, or on the average about one per Schmidt field. It is not unreasonable to suppose that, with the extension of the magnitude limit to $m_{pg} = 18$ or 19, there may be between five and ten times as many nearby stars as are known already. We are therefore planning to carry out a parallax programme on a few fields, with the UK Schmidt, in which every star will be measured.

Even with a GALAXY machine this is a major undertaking, but is just feasible in high latitude fields. Assuming a density of 1000 stars per sq deg a parallax series could be measured in about three months. No doubt, faster measuring machine will become available in due course, so that it would be worthwhile to take plates on fields with a higher star density for future parallax and ultimately proper motion measurements. The aim would be to achieve a standard error of a single parallax of say $0''.01$; a few spurious large parallaxes might then be expected on purely statistical grounds. But with a field of say 30 000 stars there should be only two with errors four times the standard error, and it should be possible to pick out stars which are really within 20 pc.

With such a programme we will not only be able to find individual stars with large parallaxes, which can then be observed on a larger telescope, it will also be possible to measure statistical trigonometric parallaxes for groups of faint stars, such as M-dwarfs out to much larger distances.

3. Proper Motions

I would like to consider proper motion work in two categories, namely surveys for stars with large motions and investigations which demand high resolution. The former is well represented by Luyten's current survey. The whole northern sky is covered by Palomar Schmidt plates at two epochs and the labour is now that of blinking, aided by automatic measuring equipment, computation and publication. However, although the southern sky has been almost completely covered to $m_{pg} \simeq 16$ by the Bruce Proper Motion Survey, the search for large motions among fainter stars must await the completion and later repetition of the southern surveys now being undertaken on the ESO and UK Schmidt telescopes.

The second category, of high resolution proper motion work, includes such problems as cluster membership and studies of stellar population, density and kinematics from proper motion dispersions, which require relative proper motions of high accuracy, and also problems requiring absolute motions such as secular parallaxes and general studies of the systematic motions in the Galaxy.

The existence of the Palomar Survey plates, or, for practical purposes, glass negative copies of them, ensures that astrometric information on about the same scale as the 'Carte du Ciel' is potentially available at epoch around 1950, for objects brighter than $m_{pg} = 21$. This material could be of great use in deriving proper motions of individual faint objects which may become of interest, and also for general kinematic studies. In particular it is in principle possible to refer such measurements directly to the system of external galaxies except in the zone of avoidance. The small scale of the Palomar survey is a disadvantage in high resolution work, nevertheless glass copies of the originals combined with modern plates have been used successfully to study proper motions in the region of the Pleiades by B. F. Jones (1973).

It seems evident that good plates taken at the fast foci of large telescopes on star clusters will always be potentially of value for membership studies of intrinsically faint stars, white dwarfs, lower main sequence stars and young stars contracting on to the main sequence. Van Maanen's cluster plates, taken at the Cassegrain focus of the Mount Wilson 60-in. have proved very useful down to about $m_{pg} = 17$, but there is an obvious need to extend the material to fainter magnitudes.

A field in which I am personally particularly interested at the present time is the study of stellar kinematics and space density from proper motions in the Selected Areas. It is very fortunate that old astrometric plates on the northern areas were taken at the Radcliffe Observatory at Oxford early in this century; these plates reach about $m_{pg} = 15$ and are providing very accurate material to this limit. However, there are several interesting questions at fainter magnitudes, for example the suspected high density of low velocity M-dwarfs near the Sun, which can only be answered by measuring proper motions. I would very much like to see an accumulation of astrometric plates down to say $m_{pg} = 20$ in the Selected Areas.

4. Conclusion

In this review I have suggested, for the purpose of discussion, various parallax and proper motion programmes, which require large telescopes. A great lead has been given by the USNO 61-in. reflector, but it is unrealistic to suggest that there should be many such telescopes primarily dedicated to astrometry. There is much more work to be done than can be undertaken on one telescope, and indeed there are problems at magnitudes beyond the practical limit of the 61-in. Astrometric specialists should therefore seek observing time on large reflectors for their own researches; but, such telescopes being multi-purpose, it is clear that careful planning and perhaps co-ordination of programmes is needed in order to make the optimum use of the time available.

References

Cannon, R. D. and Lloyd, C.: 1970, *Monthly Notices Roy. Astron. Soc.* **150**, 279.
Greenstein, J. L. and Eggen, O. J.: 1966, *Vistas in Astronomy* **8**, 63.
Jones, B. F.: 1973, *Astron. Astrophys. Suppl.* **9**, 313.
Jones, D. H. P.: 1973, *Monthly Notices Roy. Astron. Soc.* **161**, 19P.
Luyten, W. J.: 1963, *Minnesota Publ.* **III**, No. 13.
Luyten, W. J. and La Bonte, E. A.: 1972, *The Role of Schmidt Telescopes in Astronomy*, Hamburger Sternwarte, p. 33.
Murray, C. A.: 1971, *Monthly Notices Roy. Astron. Soc.* **154**, 429.
Murray, C. A., Corben, P. M., and Allchorn, M. R.: 1965, *Roy. Obs. Bull.* **91**.
Murray, C. A. and Sanduleak, N.: 1972, *Monthly Notices Roy. Astron. Soc.* **157**, 273.
Parkes, A. G. and Penston, M. V.: 1973, *Monthly Notices Roy. Astron. Soc.* **162**, 117.
Riddle, R. K.: 1970, *Publ. USNO, 2nd Ser.* **XX**, Part IIIA.
Routly, P. M.: 1972, *Publ. USNO, 2nd Ser.* **XX**, Part VI.
Sanduleak, N.: 1964, Thesis, Case Institute of Technology, unpublished.
Strand, K. Aa.: 1962, *Astron. J.* **67**, 706.
Strand, K. Aa. and Riddle, R. K.: 1970, *Publ. USNO, 2nd Ser.* **XX**, Part IIIC.
Van Altena, W.: 1971, *Astron. J.* **76**, 932.
Worley, C. E.: 1966, *Vistas in Astronomy* **8**, 33.

DISCUSSION

Bok: It is good to hear that Mr Murray will pay special attention to Kapteyn Selected Areas in his work on the wholesale attack for the determination of individual parallaxes in a given field. I recommend that Selected Area 141 near the South Galactic Pole be given special attention.

Luyten: I was very glad to hear that you intend to try the wholesale parallax method again. I think we could save you some time if you have forty parallax plates with 30000 stars each; we could process them all in about two days.

Gliese: I am pleased to learn that you will start a parallax programme near the South Galactic Pole with the 48 in. at Siding Spring. This will help to answer the question of the frequency of red dwarfs near the Sun. In this polar region Prof. Luyten has published several hundred red stars with proper motions $\mu \gtrsim 0\rlap{.}''2$ yr^{-1} which should be dwarfs. This is the only survey including stars fainter than $m_{pg} = 17$. The problem is that nothing is known about the true space velocity distribution of the very low-luminosity red dwarfs. Adopting the velocity distribution of the McCormick stars also for these objects I have computed the number of red dwarfs with $\mu \geqslant 0\rlap{.}''20$ yr^{-1} to be expected from Luyten's luminosity function. As the number of observed stars is somewhat smaller than expected, this preliminary estimate does not fit the assumption of an extremely large number of late dwarfs in the solar neighborhood. It even does not fit Miss Weistrop's results if the ratio of dMe to dM stars varies appreciably with luminosity.

Elsmore: May I ask Mr Murray, concerning the proper motion determinations with the Schmidt at Siding Spring, with what time intervals are successive plates taken?

Murray: I would want at least 15 years. Of course the accidental errors can be reduced by taking several plates at each epoch.

Bok: I hope that in the search work for candidates for parallax programs special attention may be given to the search of specific spectral groups by the use of variations upon the Hoag-Schroeder transmission grating approach with the grating placed near the Cassegrain Focus of an *f*/8 reflector. With the new 140 to 158 in. reflectors now getting into operation one may expect to obtain data for stars to $V = 18$ (or fainter) over areas of one square degree in the sky.

Strand: The 61 in. astrometric reflector has proven not only to be excellently useful to trigonometric parallaxes, but has also proven successful for such work as photoelectric observations of close binaries and for image tube work on galaxies and globular clusters.

Blanco: Near the Galactic Poles one can readily segregate dwarf M stars for parallax programmes by obtaining objective-prism plates with emulsions sensitive to near-infrared light. Dispersions of about 1000 to 2000 or more at the atmospheric 'A' band ($\lambda 7590$), and unwidened spectra allow one to reach relatively faint stars. The spectra are not adequate for luminosity determinations, but of the stars one finds, the fainter ones are likely to be dwarfs, since a faint giant would have to be at intergalactic distances to be so observed.

Murray: I did measure proper motions for 21 of these stars, listed by Sanduleak from Warner and Swasey objective-prism plates. It is these motions which suggest the possible high density of M dwarfs near the Sun.

Eichhorn: It is well known that the 'acceleration' of the motion of a star is connected with its radial velocity and its parallax, beside the proper motion. Since we may expect that the accuracy of proper motion determinations and radial velocity determinations will increase, perhaps drastically, it will become possible to determine individual distances from measuring the changes in the proper motions of the stars. In order to do this, it might be wise to take plates of regions at intervals of, say 25 to 50 years. We must, of course, realise that this work may conceivably benefit generations of astrometrists that will live long after all of us will have been gone.

THE U.S. NAVAL OBSERVATORY PARALLAX PROGRAM

K. Aa. STRAND, R. S. HARRINGTON, and C. C. DAHN

U.S. Naval Observatory, Washington, D.C., U.S.A.

Abstract. The U.S. Naval Observatory program on trigonometric stellar parallaxes with the 61-in. astrometric reflector has been in progress since 1964. To date 201 definitive and 8 preliminary negligible parallaxes have been published, including the UBV photometry of these stars. Data for an additional 35 stars are still unpublished.

The mean error in position for an image of unit weight is 1.2 μ or 0''016, of which 0.8 μ originates from the measuring error of the automatic measuring machine.

An error of the parallax of 0''004 can usually be obtained with 40 or fewer plates and with an average parallax factor of 0.7.

A statistical investigation of the derived parallaxes shows that they are free from significant internal systematic errors.

1. Introduction

The U.S. Naval Observatory (USNO) parallax program with the 61-in. Astrometric Reflector at the USNO Flagstaff Station in Arizona has been in operation since April 1964; to date some 25000 plates have been obtained. Since the telescope and the SAMM automatic measuring machine have been described elsewhere (Strand, 1971) no further description will be made here. The thorough testing of the telescope in the initial stage (Hoag *et al.*, 1967) showed it to be a highly stable instrument, both mechanically and optically. Its operation over the past 9 years has corroborated this.

We will briefly discuss the following at this time:

(1) The results of the program.

(2) The accuracy of the astrometric data.

(3) The photometric program.

(4) An analysis of the lower main sequence and the degenerate branch in the Hertzsprung Russell Diagram.

2. The Results of the Program

Two catalogs of stellar parallaxes have been published. The first catalog (Riddle, 1970) contains 100 parallaxes, while the second catalog (Routly, 1972) has 109 parallaxes. The third catalog, unfortunately, is not as near completion as originally scheduled because of persistent unfavorable observing conditions at Flagstaff during a six months period. Beginning with the second catalog, preliminary parallaxes and proper motions have been published for stars which show near zero parallaxes. These data are based upon limited, but sufficient, plate material which has established that further effort on these stars is not likely to produce significant parallax information. This procedure seems justified since the program is aimed at obtaining parallaxes of nearby degenerate and late main sequence stars without attempting to establish statistically significant data in regard to distribution of these stars in space. In all,

Gliese, Murray, and Tucker, 'New Problems in Astrometry', 159–166. All Rights Reserved.

seven stars within 10 pc and with no previously published parallaxes have been found. Two stars with preliminary parallaxes larger than 0".2 (Riddle *et al.*, 1971; Dahn *et al.*, 1972) are currently on the program for completion as are 8 stars with preliminary parallaxes between 0".1 and 0".2.

3. The Accuracy of the Data

The mean error in position for an image on the plate of unit weight was found to be 1.2 μ or 0".016. This error depends upon the average assigned weight of the plates in the individual series and is a somewhat arbitrary value in that a systematic increase or decrease in weight would correspondingly raise or lower this value. A further investigation of the weighting scheme (which depends upon an estimate of the quality of the image, made at the time of measurement) has shown (Harrington) that all images, regardless of weight on the present weighting scheme indicate the same mean errors of position. The average absolute parallax factor in x is 0.68. Typically a parallax based upon 36 plates has a mean error of 0".004.

In addition to these conventional expressions for the internal accuracy of the parallaxes, other tests were devised (by Harrington). One test investigated the number of plates required in a series before a parallax levelled off to its final value. This was found to be approximately 25 plates. It therefore appears that the chosen number of 36 plates in a series provides sufficient margin for establishing a stable final value for a parallax.

Another test consisted of assuming a parallax and proper motion for various program stars and attempting to re-derive those values. The position of a star was calculated for each of the times plates were taken in an actual parallax series, using the adopted parallax and proper motion and adding a random error from a normal distribution with a standard deviation of 1.2 μ (equal to the positional accuracy of SAMM). From the positions thus obtained, a fictitious parallax was calculated which, in all cases, differed only slightly from the assumed parallax. This procedure was repeated 5 times for each of the stars in the second catalog. The residuals of the fictitious parallaxes, compared with the assumed parallax, were normally distributed with a mean error of 0".004, equal to the observed internal mean error of USNO parallax determinations. From this it may be concluded that the USNO parallaxes are free from serious internal systematic errors.

The present reduction program provides, in addition to the parallax and proper motion with their mean errors of the parallax star, the preliminary values of these quantities for the comparison stars as a safeguard against having a comparison star with an appreciable parallax vitiating the final value of the parallax of the star in question. No attempt has been made to use these parallaxes to establish an external mean error of the derived parallaxes, which they obviously cannot produce.

In regard to the external errors of the derived parallaxes, using the conventional method of comparing ours with those derived at other observatories. the total number of parallaxes in common is still insufficient to establish these errors (systematic as well as random errors). A very preliminary investigation of this kind (Gliese, 1972) seems to indicate that the USNO parallaxes differ by an amount of $+0\rlap{.}''002$ $\pm0\rlap{.}''003$ from the Jenkins General Catalog of Trigonometric Parallaxes and that they have an external mean error of $\pm0\rlap{.}''0085$.

4. The Photometric Program

The photometric program, instituted for the purpose of supplementing the astrometric data with photoelectrically determined magnitudes and colors, continues with both the 61-in. (J. B. Priser) and 40-in. reflectors (C. C. Dahn and H. H. Guetter) at the Flagstaff Station. However, two modifications to the original program have been made since June 1971. They are:

(1) U bandpass measures of the lower main sequence stars have been discontinued; and

(2) $V-I$ color indices on the photometric system of Kron and Mayall (1960) are

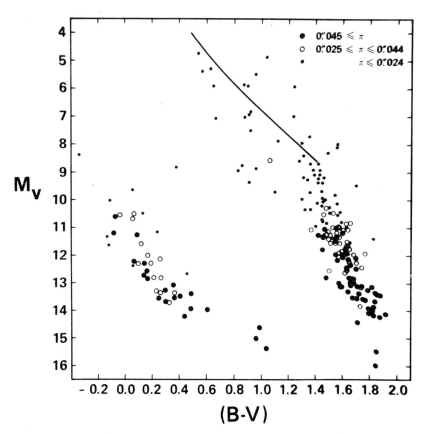

Fig. 1. The HR diagram for stars for which parallaxes and photometric data were obtained at the U.S. Naval Observatory and published in its first two catalogs.

now being measured for the red dwarf stars by Dahn and Guetter with the 40-in. Ritchey-Chrétien reflector.

The justification for making these changes is twofold:

(1) The faintness of the red dwarf program stars in the U bandpass precludes obtaining $U - B$ color indices with sufficient accuracy to warrant the effort. For example, of the approximately 70 red dwarf stars for which astrometric and photometric results were reported in the 'Second Catalog of Trigonometric Parallaxes of Faint Stars' (Routly, 1972), 60% were fainter than $U = 16.0$ and 30% were fainter than $U = 17.0$.

(2) The superiority of a color index employing a bandpass in the near infrared over one with bandpasses in the blue is well recognized for red dwarf stars.

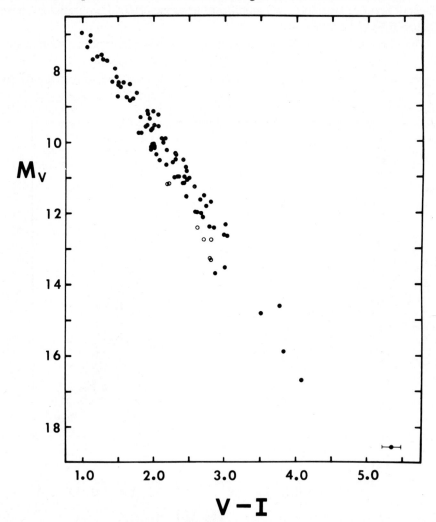

Fig. 2. M_v, $V - I$ color magnitude diagram of red dwarfs with trigonometric parallaxes larger than 0."045. Colors were determined by C. C. Dahn and H. H. Guetter at the U.S. Naval Observatory, Flagstaff Station.

5. An Analysis of the Lower Main Sequence and the Degenerate Branch in the Hertzsprung-Russell Diagram

In the Figure 1 is shown a HR diagram based entirely upon the USNO parallaxes published in the two catalogs previously referred to. The $B-V$ colors for the majority of the stars are also from the USNO program. It should be noted that the data have made a substantial contribution to the establishment of the lower main sequence and the degenerate branch.

Figure 2 shows the M_v, $V-I$ color magnitude diagram obtained in the previously mentioned $V-I$ photometry by Dahn and Guetter. The selected stars all have well determined published parallaxes, and each data point is based upon a minimum of 3 independent measures of $V-I$. Van Biesbroeck's star based upon a Kron (1958) value is included with an error bar. Open circles in the diagram identify stars classified as subdwarfs on the Mt Wilson spectroscopic system by Joy (1947).

Data obtained for 45 red stars with published USNO parallaxes larger than 0".05 are shown as crosses in the next M_v, $V-I$ diagram, Figure 3. They clearly provide a significant contribution towards delineating the lower main sequence. There are at least an additional 25 stars in the USNO program in the same category as the ones shown, for which only preliminary results are available. There are indications that about 7 of these will turn out to be intrinsically fainter than $M_v = 14$.

We shall next briefly discuss the delineation in regard to luminosity and color distribution of the Eggen (1969) subluminous stars, which he defines as having $U-V \gtrless 0.0$ and are at least 2".5 below the main sequence in the M_v vs $U-V$ color magnitude diagram. Figure 4 is reproduced from the Eggen-Greenstein (1965) paper and summarizes the observational material available at that time. The dotted line has been added to delineate the Eggen's subluminous designation. The individual stars are identified by their Lowell proper motion survey numbers.

The current status of the stars in question as determined from published parallaxes and from unpublished preliminary USNO parallaxes is shown in Figure 5. Of the 15 stars identifiable with the red degenerates, USNO parallaxes have been published for 6 and preliminary parallaxes are now available for an additional 7 of these stars.

We shall at this time only discuss the three stars lying above the red degenerate branch, G14-24, HG7-138 and G5-28. The USNO parallax for G14-24 is 0".011 $\pm$ $\pm 0".004$ (m.e.) contradicting the Cape parallax of 0".074, and supports the spectroscopic classification as being a weaklined K subdwarf (thereby permitting Greenstein and Eggen (1966) to retain their spectroscopic badges). In the same manner the very preliminary USNO parallax ($\pi = -0".008 \pm 0".009$ (m.e.)) for HG7-138 suggests it is not a member of the Hyades cluster.

G5-28 with a USNO parallax of $+0".020 \pm 0".003$ appears to definitely depart from the other wise well defined branch. (If the star should conform to the degenerate branch it should have a parallax of 0".072, while a parallax of 0".005 would place the star among the subdwarfs.) In view of the size of the observed parallax the latter is the more likely case. With this exception we conclude on the basis of the USNO data

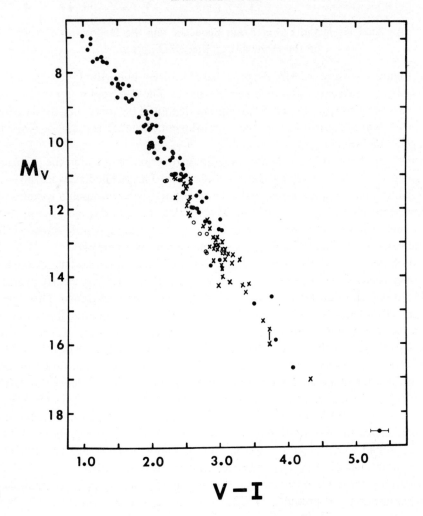

Fig. 3. M_v, $V-I$ color magnitude diagram of red dwarfs shown in Figure 2 with additional data ($\times$) obtained from U.S. Naval Observatory parallax program.

accumulated to date – data which includes near-zero parallaxes for another 10 or so stars which Eggen has suggested as good red subluminous candidates – that (1) slow but steady progress is being made in discovering new candidates for the red degenerate sequence; (2) no parallax sufficiently large to go unchallenged has yet to appear among the USNO program stars to support the contention that stars do populate the region of the M_v vs $U-V$ color-magnitude diagram between the subdwarfs and the red degenerate sequence.

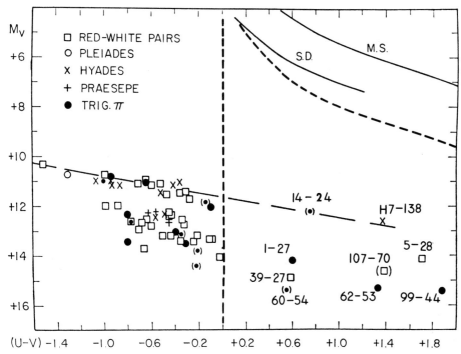

Fig. 4. M_v, $U-V$ diagram for white dwarfs, adopted from O. J. Eggen and J. L. Greenstein (*Astrophys. J.* **141**, 95, 1965).

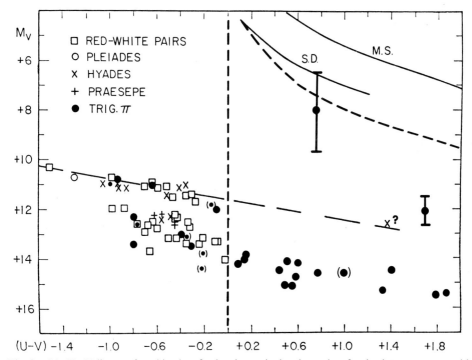

Fig. 5. M_v, $U-V$ diagram for white dwarfs, showing revised and new data for the degenerate stars with $U-V > 0.0$.

References

Dahn, C. C., Behall, A. L., Guetter, H. H., Priser, J. B., Harrington, R. S., Strand, K. Aa., and Riddle, R. K.: 1972, *Astrophys. J.* **174**, L87.

Eggen, O. J.: 1969, *Astrophys. J.* **157**, 287.

Eggen, O. J. and Greenstein, J. L.: 1965, *Astrophys. J.* **141**, 83.

Gliese, W.: 1972, *Quart. J. Roy. Astron. Soc.* **13**, 138.

Greenstein, J. L. and Eggen, O. J.: 1966, *Vistas in Astronomy* **8**, 63.

Hoag, A. A., Priser, J. B., Riddle, R. K., and Christy, J. W.: 1967, *Publ. USNO, 2nd Ser.* **XX**, Part II.

Joy, A. H.: 1947, *Astrophys. J.* **105**, 96.

Kron, G. E.: 1958, *Publ. Astron. Soc. Pacific* **70**, 102.

Kron, G. E. and Mayall, N. U.: 1960, *Astron. J.* **65**, 581.

Riddle, R. K.: 1970, *Publ. USNO, 2nd Ser.* **XX**, Part IIIA.

Riddle, R. K., Behall, A. L., Guetter, H. H., and Christy, J. W.: 1971, *Publ. Astron. Soc. Pacific* **83**, 210.

Routly, P. M.: 1972, *Publ. USNO, 2nd Ser.* **XX**, Part VI.

Strand, K. Aa.: 1971, *Publ. USNO, 2nd Ser.* **XX**, Part I.

DISCUSSION

Harris: You said that the parallax determination levels out after 25 plates; does this also apply to the internal error?

Strand: Yes, but larger than the final value.

Dieckvoss: Do you take double plates against emulsion shift as the refractor people do?

Strand: No, such a procedure would complicate the observing schedule. Since the telescope operation is practically automated the observer is ready with the next plate as soon as the previous plate is completed.

I should mention that the automatic guider guides directly on the polar and declination axes, thus leaving the plate stationary with respect to the optical axis.

Franz: Since a large proper motion is a necessary but not sufficient characteristic for selecting nearby M-dwarfs, would it not be desirable to determine first $V - I$ colours and then, on the basis of proper motions and colours, place suitable stars on a parallax programme. It obviously takes much less time to determine colours than it takes to try to measure (unmeasurable) parallaxes.

Strand: This is the procedure by which M-dwarfs are now added to the programme.

Van Altena: (1) A comparison of 17 stars in common with the Yerkes and USNO parallax programmes yields a systematic error of $+ 0\overset{''}{.}0006$ in the sense Yerkes – USNO. The difference in the parallaxes yields an external error 40% larger than the internal errors if both errors for both observatories are increased by the same amount.

(2) The bifurcation of the white dwarf sequence in the M_v vs $U - V$ diagram is a peculiarity of the photometric system. The Stromgren four-color system yields a single well-defined sequence, as shown by Graham and Weidemann.

Strand: Our results indicate that there is also no bifurcation in the $U - V$ diagram for the white dwarfs for $U - V < 0.0$.

REPORT ON THE HERSTMONCEUX I.N.T.
PARALLAX PROGRAMME

B. F. JONES

Royal Greenwich Observatory, Herstmonceux, England

Abstract. In 1971 a pilot parallax programme was initiated on the 98-in. Isaac Newton Telescope at the Royal Greenwich Observatory. Nineteen stars between visual magnitude 15 and 19 and of high proper motion were put on the programme. The plates were taken at prime focus through the Wynne corrector. Observations are now complete for two of these stars. Because of time allocations and the desire to observe at maximum parallax factor, observations are made at large hour angles. All stars in each field are being measured on the GALAXY measuring engine and parallaxes and proper motions will be computed for all stars. The fields are being reduced using an overlap reduction technique.

DISCUSSION

Bok: May I make a special plea for including certified faint members of the Hyades on all parallax programmes with large telescopes?

THE PALOMAR PROPER MOTION SURVEY

W. J. LUYTEN

University of Minnesota, Minneapolis, Minn., U.S.A.

Abstract. The present state of the Palomar Proper Motion Survey is described. Before the CDC automated-computerized plate scanner became operational in 1972, 153 plate pairs had been blinked, whereas since then 443 pairs have been machine processed. Of the remaining 350 fields, some 40 are probably beyond the capability of the present system due to excessive star density.

Some 21 000 new motions larger than $0\rlap{.}''18$ annually have been published. The accuracy of an individual motion is about $\pm 0\rlap{.}''016$ from a single machine-processed plate pair, as compared with $\pm 0\rlap{.}''025$ from hand measured plates.

There is an indication that the number of large motions per unit area increases with galactic latitude.

The National Geographic-Palomar Observatory Sky Survey consists of 936 pairs of blue and red plates taken with the 48-in. Schmidt telescope. They completely cover the sky from the North Pole to declination $-33°$ (1875), or 77% of the entire sky, and generally go down to about $m = 20$ (red) and 21.2 (blue). Since near this limit stars with sizable proper motions are not likely to be white it is obvious that for a proper motion survey the red plates must be repeated.

After a preliminary pilot program in 1958, the systematic Palomar Proper Motion Survey was begun in September 1962. First of all I should like to pay tribute to the Hale Observatories for their generosity in awarding guest-investigator privileges to me for the nine years 1962–1971. Altogether I have had twenty-five such assignments of dark-of-the-Moon nights with the 48-in. telescope, and this totals up to an equivalent of ten months use of this marvelous telescope. This enabled me to retake virtually all of the red plates, though some seventy of them were taken by the Observatory staff and loaned to me.

Financially, the program was made possible by continuous grants from the National Science Foundation for the next eleven years while the original acquisition of a complete set of duplicate negatives of the Survey was achieved through a grant from the Hill Family Foundation.

In 1965 I received a contract from the National Aeronautics and Space Administration, with Control Data Corporation as subcontractor to fabricate an automated-computerized plate scanner and measuring machine. This was completed in 1970, and to celebrate the event we held an IAU Colloquium on Proper Motions in Minneapolis in April of that year. Owing to various delays of different nature we didn't really get going with the processing of plates until January 1972. In the nineteen months since that time the machine has proved to be spectacularly and fantastically successful and since I personally had little or nothing to do with the design or the construction of it, I can say that, in my opinion, this plate scanner constitutes the most important addition to our astronomical instrumentarium in the last twenty-five years – with the exception of the big optical telescopes and radio dishes. I should like

Gliese, Murray, and Tucker, 'New Problems in Astrometry', 169–173. All Rights Reserved.

to express my deep appreciation to the engineers of Control Data Corporation, and especially to James Newcomb, who was largely responsible for the hardware, and the optical system, and to Anton La Bonte who designed the software, and who guided us through the crucial periods of 'debugging', when the machine still exhibited various 'children's diseases'.

And so now, for the first time, I can really say – not what we are going to do, but what we actually have accomplished.

Between 1962 and 1971 we hand-blinked 153 pairs of plates on which we found 77000 proper motion stars. These have all been measured, and data for 62000 of them have been published. The remainder are ready for publication and should appear soon.

As of the day I left Minneapolis we had processed 443 pairs of plates with the machine, but this included ten pairs which had been previously hand-blinked.

At present the count thus stands as follows:

Hand-blinked	153	
Machine-processed	443	
(which includes	10	previously hand-blinked)
Number of fields processed	586	
Remaining	350	
	936	

Of the 350 pairs still remaining 117 can be processed now, and we hope to complete these by the end of 1973. The final 233 pairs are low-galactic latitude plates with very high star densities and these will require some changes in the software before they can be processed. We feel pretty confident that we can handle some 150 of them, and perhaps close to 200, but probably some 40 pairs are beyond the capabilities of the scanner at the moment.

Incidentally, since the number of star images appearing on the plates averages at least 80000 it means that in processing 443 pairs of plates we have determined x and y coordinates to one tenth of a micron for at least 70 million stars.

In publishing our data there is some difference in what we give as between the hand-blinked and the machine-processed plates. For the hand-blinked plates we give positions to $0\overset{m}{.}1$ in R.A. and $1'$ in Decl.; we give estimated photographic magnitudes and colors and motions to $0\overset{''}{.}01$ and $1°$. For the machine-processed plates we have computer print-outs giving R.A. to $0\overset{s}{.}1$, Decl. to $1''$ (though we publish only to 1^s and $0\overset{''}{.}1$), machine-determined red magnitudes and proper motions to $0\overset{''}{.}001$ and $1°$. We now have such data for close to a quarter of a million stars with motions larger than $0\overset{''}{.}09$ annually. Obviously, most of these data are of statistical interest only, but the stars with the larger motions merit individual attention. Since all our motions are relative and require corrections to render them absolute, and since in extreme cases this correction might amount to $0\overset{''}{.}016$, I have arbitrarily selected $0\overset{''}{.}18$ annually as the lower limit of motion for stars of individual interest – this to make sure that no motion larger than 0".2 annually would be overlooked. Thus, after the first print-out has been obtained, we run off a second one which lists, first, all motions larger than 0".18 annually, and, second all motions between 0".18 and 0".09. For every star in the first list I personally check the motion and estimate the photographic magnitude and the color from the blue and red plates.

These data are published as rapidly as possible, and to date we have published about 14000 new motions larger than 0".18 which, together with the hand-blinked plates gives us some 21000 new motions larger than 0".18 annually from the Palomar plates alone. This is more than twice as many as have been published by all other proper motion surveys together – except of course, for the Bruce Survey. When the entire Survey has been completed we hope to publish a catalogue of more than 50000 stars with motions larger than 0".2 annually. We now also have more than 1000 new motions larger than 0".5 annually from the Palomar plates alone, and another 1000 low-luminosity stars from the machine-processed plates. Further we have another 1500 new white dwarfs, for a total of some 4500, and 1600 new double stars with common proper motion, bringing our total to 4700, and these include some 250 binaries with one degenerate component, about ten with both components degenerate, and two with both components virtually identical white dwarfs. Among these 1600 are a few triples but these amount to fewer than one percent – and to-date we have not yet found even one genuine quadruple and certainly no new moving clusters.

It is not feasible to check individually the more than 200000 smaller motions, which possess mainly statistical value. The data for these will probably not be published in book-form, but will be provided to, and made available at the NASA Data Center at Greenbelt Maryland, either on microfilm, or on magnetic tape, or both. For all of these stars we shall also provide a Quality-index-rank, based on several quality indices devised by La Bonte which depend on the roundness of the image and several other factors, and we shall provide statistical probabilities for and against the reality of the motions of any given quality rank. For these stars we have only machine-determined red magnitudes and no colors.

Two further questions need to be answered viz. how complete is our survey, and what is the accuracy of the motions. We have no real data on the completeness as yet – only guesses. We know, e.g. that the computer rejects double stars, especially in the 4–10″ separation range when the components are nearly equal, because the images are not round. We have also programmed out all motions larger than 2".5 annually and most stars with images larger than 250 μ diameter because of the uncertainties in the reconstruction of the image-centers, though we need some of these, as SAO stars, for the determination of celestial positions. Our present guess is that for motions between 0".2 and 2".5 annually we are at least 90% complete for stars brighter than 18 (red) but we do not really know whether possible, and occasional mal-functioning of the machine can reduce this drastically in some areas.

As to the accuracy we have some real data here. The hand-blinked, and hand-measured plates give motions with a mean error of about 0".025 – this is determined from overlaps, and by comparison with previous Bruce motions. For the machine-processed plates we find, from overlaps between adjacent plates, that the rms error for a motion measured on a single pair of plates is about $0''\!.016$ for stars fainter than 12.5 (red), if the plate interval is 11 years or more. For brighter stars the errors are undoubtedly larger, and this is also true in the few instances where we were forced to process plates with intervals much shorter than 11 years.

From the machine-processed plates there emerges a definite indication that the number of large motions per unit area decreases markedly as we go from the galactic poles to lower galactic latitudes. It is, of course, conceivable that, as we go to lower latitudes, and higher star-densities our machine processing becomes progressively more incomplete. However, within the range of magnitude and motion accepted by the computer we seem to pick up roughly the same percentage of known motions and I am inclined, therefore, to accept this decrease in the frequency of motions with latitude as at least partially real.

Finally, a few comments on the first large area which has been completely processed, viz. the region of the South Galactic Pole. As boundaries we took 22^h and 3^h40^m in R.A. and $+4°$ to the southern plate edge – but this varies from $-32°53'$ to $-32°42'$ (1950) because of precession. Within this area, which totals some 3000 sq deg or 0.07 of the sky we list a total of 6970 stars from all sources, brighter than 21.5 pg and with motions larger than $0''\!.18$ annually. Of these, 5900, or 86% were found in the Bruce or Palomar Surveys. Because of this large number of stars in a restricted area it has become possible, for the first time, to analyze the magnitude distribution for groups of stars with total proper motions between very narrow limits, which means, stars which are statistically at the same distance. In this way, then, one could obtain a new and independent check on where the maximum of the luminosity function lies. The details are given in the publication on the South Galactic Pole, and here I need only mention the result which comes to:

Maximum of the luminosity function lies at $M = +15.4\ pg$.

We hope to do the same for the North Galactic Pole region and here we are not limited in declination to 33° as we were at the South Galactic Pole, and we expect to analyze the entire area between 10^h and 15^h40^m, and $-4°$ and $+57°$ or about 4400 sq deg or 0.107 of the sky. This region is covered by 130 plates all of which have now been hand-blinked or machine processed, and we expect to make up a final catalogue of some 10000 stars with motions larger than $0''\!.18$ annually and brighter than 21.5 pg, for all of which we shall have colors. We hope, also, to obtain statistical information on a further 50000 stars with motions between $0''\!.18$ and $0''\!.09$ but we shall have only red magnitudes and no colors for the majority of these.

DISCUSSION

Klock: To what precisions may the magnitude be determined by your machine?

Luyten: The repeatability of the machine measurement on one plate gives a mean error of about $0^m\!.15$ – on overlaps of different plates it is probably a little larger than $0^m\!.2$.

Franz: How close can the images of two faint stars be on a Palomar Schmidt plate and still be measured separately with normal accuracy?

Luyten: This is really the main flaw in our present procedure. Double stars with equal components and separations of 4″–10″ are likely to be rejected by the computer but more narrow and wider doubles are generally accepted. Doubles with a large magnitude difference generally are accepted as a single star.

Van Herk: Is the number of stars with large proper motions increasing from the galactic plane towards the poles?

Luyten: Definitely – the only reservation I have is that possibly, in the lower galactic latitudes where the star-density is greater, the machine misses more proper motion stars. However the percentage of known proper motion stars which the machine is expected to pick up appears to be the same in high as in low latitudes.

Eichhorn: This machine is obviously an extremely useful piece of equipment, and potentially of enormous value for photographic astrometry. Can you comment on the future of the machine, in particular, make a statement as to whether there are any plans to deposit the machine – at least after the termination of your project – at a place where it would be accessible to other qualified observers, at least in the U.S.?

Luyten: If all goes well we expect to finish the processing of all high latitude plates this calendar year. The high-density low-latitude plates will require changes in the software, and this may take a year or more. But we should be completely finished sometime in 1975 and personally I should welcome applications for the use of the machine. It is an extremely versatile instrument and it is not expensive to operate

Fricke: Could Dr Luyten tell us something about the faint end of his improved luminosity function, in particular, to what absolute magnitude limit is the function reliably known in his opinion?

Luyten: I have only statistical evidence, as there are no parallaxes. The faintest star we have found is 20.8 m_{pg} with a motion of 1″77. I think it is safe to say that down to absolute magnitude 17 m_{pg} we are reasonably sure, at 18 we are getting incomplete and at 19 we end.

ACCURATE OPTICAL POSITIONS USING THE PALOMAR SKY ATLAS

R. W. HUNSTEAD

Cornell-Sydney University Astronomy Centre, School of Physics, University of Sydney, Sydney, Australia

Abstract. Optical positions measured directly on the prints and plates of the Palomar Sky Atlas have been compared with various published lists of precise optical positions. The external comparisons suggest that positional errors $\sim \pm 0\rlap{.}''3$ in each coordinate have been achieved. This supports the claimed accuracy and establishes the Sky Atlas as an important astrometric medium for investigating radio-optical associations.

1. Introduction

The measurement of precise optical positions for the objects associated with radio sources presents some special astrometric problems. The most important of these results from the shape of the radio luminosity function: the strongest radio sources tend also to be the faintest optically, being mostly fainter than 17^m. This effectively rules out the conventional astrograph for routine identification and astrometric work. The Sky Atlas prepared from the Palomar Observatory Sky Survey (POSS) has therefore been used almost exclusively as the initial search medium for optical identifications, because it reaches faint magnitudes (20^m–21^m) and also offers colour information. However, the view has been widely held that the Sky Atlas prints are quite unsuitable for measuring positions to better than a few arc seconds due to distortions of the paper base (Dewhirst, 1963). Consequently, the much-needed astrometry continues to be carried out on Schmidt plates specially taken for this purpose at Palomar, Cambridge and Asiago.

The aim of the present paper is to examine in detail the accuracy of direct measurements on the second-generation copies of the POSS in relation to published lists of optical positions obtained from original plate material. The POSS position measurements upon which the following analysis is based have been reported in an earlier paper (Hunstead, 1971), together with a description of the measuring engine and the measurement and reduction techniques.

2. Comparisons with Other Optical Positions

The most general method of assessing the external accuracy of different sets of measurements is to evaluate unweighted variances about a mean difference for each comparison. Systematic differences will be influenced primarily by regional fluctuations amongst the different star catalogues and a discussion of these effects will be deferred until a later paper.

The six lists of optical positions used in the present comparison are given below with the reference star catalogues used.

Gliese, Murray, and Tucker, 'New Problems in Astrometry', 175–179. All Rights Reserved.

Observers	Abbreviation	Catalogue
(i) Argue and Kenworthy (1972)		AGK3
Argue *et al.* (1973)	AK	AGK3, SAO
(ii) Barbieri *et al.* (1971)		SAO
Barbieri *et al.* (1972)	BCGP	AGK3, SAO
(iii) Bolton (1968)	B	Yale
(iv) Hunstead (1971)	H	SAO
(v) Kristian and Sandage (1970)	KS	AGK3, SAO
(vi) Véron (1956a, b, 1966, 1968);		
Bolton *et al.* (1965); Sandage *et al.* (1965)	V	AGK2, SAO

With the exception of (iii) and (iv), the positions were mostly referred to AGK north of $-2°$ and SAO southwards.

The results of inter-comparisons amongst sources common to each pair of lists are summarised in Table I, which gives the (unweighted) standard deviation σ' about a mean difference for each comparison $(\sigma' = \frac{1}{2}(\sigma_\alpha + \sigma_\delta))$. The diagonal entries in the table correspond to the external accuracies claimed for each list.

TABLE I

Standard deviations (mean of R.A. and Decl.) in a single position difference amongst various sets of optical position measurements. The underlined entries along the diagonal give the external errors estimated by each observer. A dash (–) indicates fewer than five objects in common.

	AK	BCGP	B	H	KS	V
AK	0″2	0″68	–	0″34	0″30	1″0
BCGP		0″4	1″2	0″75	0″70	1″3
B			0″5	1″1	1″3	–
H				0″4	0″38	1″1
KS					0″3	1″0
V						1″0

Despite the fact that a different number (always $\geqslant 6$) and selection of sources were used in each comparison, the data in Table I give a surprisingly consistent picture of the intrinsic accuracy of each list of positions. Only in the cases of B and BCGP do the actual errors appear to be underestimated. Most important, however, are the comparisons amongst AK, KS and H which suggest that the direct measurements on the POSS are statistically accurate at the 0″3 level. It is worthwhile placing the information of Table I in better perspective by considering the intrinsic accuracy (σ) of each list in terms of the observer effort. We may regard $1/\sigma$ as measuring the 'return' on an 'investment' of N measurements, where N is defined here as the mean number of separate **x**, **y** measurements for a concluded position. The quantity $E = 100/\sigma N$ therefore provides an 'efficiency measure' for each list of positions. Figure 1 shows σ plotted as a function of N, with lines corresponding to isophotes of E.

The information in Figure 1 may be interpreted in several ways. From the viewpoint of efficiency and convenience alone, the present technique using the Sky Survey is clearly superior. However, there are also situations that appear to demand the most precise optical positions, such as,

(a) defining zero-point corrections for precise radio measurements when the number of calibrators is small (Smith, 1971); and

(b) the more general problem of investigating radio-optical displacements for the most compact radio sources.

The former problem can obviously be overcome by choosing a larger number of calibrators, although this may not always be practicable for radio telescopes with

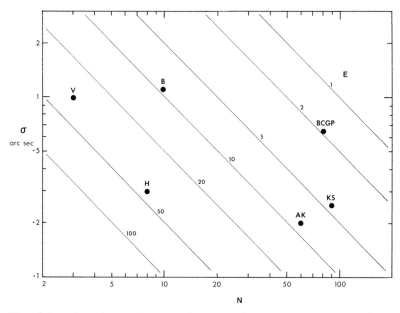

Fig. 1.   Plot of the estimated external error, σ (arc seconds), against the number, N, of separate x, y measurements for a concluded position in each of the six optical lists. The lines correspond to isophotes of 'efficiency measure', E, where $E = 100/\sigma N$.

beams of the order of a few arc seconds; the problem should nevertheless be only a temporary one. The degree to which the second problem can be pursued will probably be determined primarily by the limitations of conventional photographic astrometry, i.e. measurement errors, plate distortions, photographic noise and 'warps' in the fundamental star catalogues. At present it would seem to be more profitable astrophysically to direct the optical effort towards an investigation of spatial relationships between optical sources and radio structures on the scale of a few arc seconds. For this purpose, astrometry using the Sky Atlas would permit an extensive quantitative study of the physics of non-thermal sources, as well as enabling reliable identifications to be made at a rate far greater than is feasible with specially taken plates. A further advantage is that this work may be undertaken by the radio astronomers themselves.

3. Outlook

The question now arises as to whether it is possible to improve significantly on the $0\rlap{.}{''}3$ figure derived earlier for the POSS measurements. The answer is almost certainly no – not without eroding the present advantage in efficiency. An analysis of repeated measurements gives an internal error of $0\rlap{.}{''}27$ rms in each coordinate while the uncertainty of fit to SAO will be $\sim 0\rlap{.}{''}3/\sqrt{8} = 0\rlap{.}{''}11$ rms; combining these gives an estimated external error of $0\rlap{.}{''}29$, in close agreement with the value of $0\rlap{.}{''}30$ deduced from the earlier comparisons. The $0\rlap{.}{''}27$ internal error can be attributed mainly to photographic noise at the QSO or galaxy level affecting the setting accuracy. This value is surprisingly small when account is taken of the variability in image quality from print to print and between the centre and edge of each print, quite apart from the overall deterioration resulting from the two additional photographic processes. The use of fine-grained IIIa-J emulsions for the southern sky survey to be commenced shortly by the British 48-in. Schmidt telescope may be expected to give a reduction in this noise component.

A further point worth mentioning concerns the star catalogues presently available for the southern sky. A major portion of the positional errors in the SAO at epoch 1950 arise from propagated uncertainties in the stellar proper motions (Dieckvoss, 1963). The second-epoch plates for the Southern Yale Zone Catalogues, which contribute more than 50% of the SAO stars south of the equator, were taken in 1933. Since most of the POSS was carried out between 1950 and 1955, it is clear that on average the random and systematic errors in an SAO reference frame defined on the POSS will be only $\frac{1}{2}-\frac{2}{3}$ as serious as those on recent plates using the same star field. This in turn means that an SAO reference frame specified by 8 stars with a favourable distribution on the POSS plates is more precisely defined than one specified by ~ 20 standards on a recent plate. The residuals quoted by Barbieri *et al.* (1972) tend to support this argument.

4. Conclusions

The principal advantage of the POSS lies in its demonstrably greater efficiency in securing unambiguous radio-optical associations. It is clear from the foregoing analysis that direct position measurements on the POSS give adequate precision for all but the most demanding situations. In the latter case, it is arguable whether a uniform, absolute accuracy of $0\rlap{.}{''}1$ or better can in fact be achieved with conventional optical astrometry to match the anticipated precision of radio astrometry.

Acknowledgements

This work has been supported by the Australian Research Grants Committee, The Sydney University Research Grants Committee, and the Science Foundation for Physics within the University of Sydney.

References

Argue, A. N. and Kenworthy, C. M.: 1972, *Monthly Notices Roy. Astron. Soc.* **160**, 197.
Argue, A. N., Kenworthy, C. M., and Stewart, P. M.: 1973, *Astrophys. Letters* **14**, 99.
Barbieri, C., Capaccioli, M., and Pinto, G.: 1971, *Contrib. Obs. Astrofis. Asiago*, No. 245.
Barbieri, C., Capaccioli, M., Ganz, R., and Pinto, G.: 1972, *Astron. J.* **77**, 444.
Bolton, J. G.: 1968, *Publ. Astron. Soc. Pacific* **80**, 5.
Bolton, J. G., Clarke, M. E., Sandage, A. R., and Véron, P.: 1965, *Astrophys. J.* **142**, 1289.
Dewhirst, D. W.: 1963, in H. P. Palmer, R. D. Davies, and M. I. Large (eds.), *Radio Astronomy Today*, Harvard University Press, p. 178.
Dieckvoss, W.: 1963, *Stars and Stellar Systems* **3**, 40.
Hunstead, R. W.: 1971, *Monthly Notices Roy. Astron. Soc.* **152**, 277.
Kristian, J. and Sandage, A.: 1970, *Astrophys. J.* **162**, 391.
Sandage, A., Véron, P., and Wyndham, J. D.: 1965, *Astrophys. J.* **142**, 1307.
Smith, J. W.: 1971, *Nature Phys. Sci.* **232**, 150.
Véron, P.: 1965a, *Astrophys. J.* **141**, 332.
Véron, P.: 1965b, *Astrophys. J.* **141**, 1284.
Véron, P.: 1966, *Astrophys. J.* **144**, 861.
Véron, P.: 1968, *Ann. Astrophys.* **31**, 483.

DISCUSSION

Dieckvoss: It is better to include 3rd order distortion of $0\overset{''}{.}36/1°$. At Bergedorf diapositives from sky survey charts are measured with the 2-screw Mann machine. Carte du Ciel charts may also be measured (to 14th limiting mag.).

Hunstead: The Schmidt bending correction is very small in comparison with the stretching effects on the Sky Survey prints. However, it is a systematic effect and maybe should be included.

Luyten: Our own experience agrees with what you have stated: we find that with our automatic measuring machine we get mean errors for a single image measurement on one plate of $\pm 0\overset{''}{.}15$ from overlapping plates.

Eichhorn: I don't want to argue whether it is desirable to go to higher accuracy than that corresponding to $0\overset{''}{.}4$ or $0\overset{''}{.}3$ standard error, but more sophisticated astrometric methods can certainly achieve higher accuracy, although probably not on the Palomar Schmidt paper prints.

Hunstead: I was talking about objects generally fainter than 18th mag.

Elsmore: In confirmation of Dr Hunstead's claims, I would like to report that at the Institute of Astronomy, Cambridge C. Hazard and A. N. Argue have made a similar investigation and find that very similar accuracies to those you report can be achieved, working from the prints.

Van Altena: How close are the reference stars to the radio source, and do you encounter much paper stretch?

Hunstead: The reference stars are always within ~ 0.9 degrees due to the limitations of the measuring engine.

We do find paper stretch but it appears to be very uniform from print to print.

Wall: With Dr Hunstead's kind co-operation, a version of the simple measuring engine in the Department of Astrophysics, Sydney University, was constructed at the Parkes Observatory. Subsequently, measurements of a number of optical positions of radio sources have been made from the Parkes copies of prints of the Palomar Sky Survey. Setting residuals are $\pm 2\ \mu\text{m}$, similar to those found by Dr. Hunstead for the Sydney University measurements, while the mean residuals for the reference stars (taken from the SAO catalogue) are $\sim 0\overset{''}{.}5$. These mean residuals are slightly larger than those found by Dr Hunstead; the difference may be due to the controlled environment in which the Sydney University prints are kept.

REDUCTION OF SCHMIDT PHOTOGRAPHS WITH THE AID OF REFERENCE MARKS ON THE CURVED PLATE

P. LACROUTE

Strasbourg Observatory, France

Abstract. A method is proposed for improving the astrometric accuracy obtainable from curved-field Schmidt plates. A reseau of small crosses in a 10 mm-spaced rectangular grid is imposed on the plate while still curved in its holder. The reseau is calibrated by photography of a field of reference stars with known positions. The method eliminates the troublesome errors arising from uncertainty in the nature of deformation of the curved plates.

DISCUSSION

Dieckvoss: The whole field of the Bergedorf Schmidt cannot be used. I recommend a device such as Prof. Lacroute's for studying the irregular bending effects.

Lacroute: I, too, would like to try to improve it by using marks.

Gliese, Murray, and Tucker, 'New Problems in Astrometry', 181. All Rights Reserved.
Copyright © 1974 by the IAU.

DISCUSSION AFTER SESSION D

Fricke: How can one improve the accuracy of positions referred to reference stars from the measurements of the Palomar Survey?

Murray: In answer to Prof. Fricke I would say that more plates should be taken and measured in the Zone Surveys. The resolution of AGK3 is currently about 0″.3 in each co-ordinate per star.

Stoy: The ±0″.3 quoted by Dr Hunstead probably reflects merely the errors in the Yale places of the reference stars, epoch 1933 brought forward to 1955 for the Sky Survey.

Improvement of this demands more photographic astrometric work.

Hunstead: Both the formal calculations and the results from repeat measurements suggest that it is the setting error on faint objects that dominates, not the catalogue positions.

Lacroute: I agree on the improvement of reference stars to improve the solutions. But if my trial is successful, that will be a contribution to the improvement of the solution by increasing the effective number of the reference stars.

Gliese, Murray, and Tucker, 'New Problems in Astrometry', 183. All Rights Reserved.
Copyright © 1974 by the IAU.

PROPER MOTIONS AND GALACTIC PROBLEMS

REMARKS ON THE DEFINITION AND DETERMINATION
OF PROPER MOTIONS

(Invited Paper)

W. DIECKVOSS

Hamburger Sternwarte, F.R. Germany

Abstract. Classical methods of compiling catalogues of proper motions from catalogues of positions at different epochs and reduction to a rigorously defined homogeneous coordinate-system are reviewed, and the determination of proper motions of numerous stars by photographic differential methods, partly with assumptions on behaviour of faint reference stars, and partly by attachment to reference frames of stars with proper motions known from catalogues is discussed.

1. The proper motion of a star is defined as the time rate of change of its position on the celestial sphere. The proper motion corresponds to the apparent transverse velocity in a fixed coordinate system, or a projection of the true velocity onto the tangential plane. The effect of the usual procedure is that the time variable is expressed in ephemeris time, because the epochs given for positions are calculated from the beginning of the relevant Besselian year, which in turn is linked with the motion of the Earth around the Sun. Except for Barnard's star, Groombridge 1830, and perhaps one or two other stars, the motion of a star may always be represented by a uniform motion in a great circle. Strictly speaking this applies to the centre of gravity in the case of a multiple system. It is only for very nearby stars with appreciable radial velocity that the second order time derivative of position enters, which gives the opportunity of checking astrometric data.

By representing the proper motion as apparent velocity in a great circle, say, in seconds of arc per century (with current precision, it is not necessary to distinguish between Ephemeris, Tropical or Julian centuries), and by specifying the position angle of the motion in a fixed equatorial system, it is possible to avoid singularities near the celestial poles. The two positions $\alpha_1, \delta_1; \alpha_2, \delta_2$ for the two epochs t_1, t_2 are interconnected by the following formulae, where $\Delta\alpha = \alpha_2 - \alpha_1$, the displacement is expressed by w, and the position-angle by P:

$$
\begin{aligned}
\sin w \ \sin P &= \cos\delta_2 \ \sin\Delta\alpha \\
\sin w \ \cos P &= \cos\delta_1 \ \sin\delta_2 - \sin\delta_1 \ \cos\delta_2 \ \cos\Delta\alpha \\
\cos w \ \ \ \ &= \sin\delta_1 \ \sin\delta_2 + \cos\delta_1 \ \cos\delta_2 \ \cos\Delta\alpha
\end{aligned}
\tag{1}
$$

and

$$
\begin{aligned}
\cos\delta_2 \ \sin\Delta\alpha &= \sin P \ \sin w \\
\cos\delta_2 \ \cos\Delta\alpha &= \cos\delta_1 \ \cos w - \sin\delta_1 \ \sin w \ \cos P \\
\sin\delta_2 \ \ \ \ &= \sin\delta_1 \ \cos w + \cos\delta_1 \ \sin w \ \cos P
\end{aligned}
\tag{2}
$$

Gliese, Murray, and Tucker, 'New Problems in Astrometry', 187–191. All Rights Reserved.

When w is small in (1) one may substitute

$$\cos \Delta\alpha \sim 1, \quad \sin \Delta\alpha \sim \Delta\alpha, \quad \sin w \sim w$$

and one gets the usual approximation in the tangential plane:

$$\begin{aligned} w \sin P &= \Delta\alpha \cos\delta_2 \\ w \cos P &= \delta_2 - \delta_1. \end{aligned} \tag{3}$$

If μ stands for the total proper motion, then $w = \mu(t_2 - t_1)$.

2. The procedure of deriving proper motion from positions at different epochs seems to be straightforward. In detail there are several complications involved. The most important will be the necessity of creating a common coordinate system conforming to an established fundamental system of positions and proper motions. At this point the real difficulties show up. Change on account of precession in the case of different equinoxes of the two catalogues is simple as long as there is a consensus on the precessional constants. But it often becomes necessary to take into account changes of constants between the time when the two catalogues were being prepared for publication. Thus if somebody makes excerpts from the Geschichte des Fixstern-himmels he has to bear in mind that here the reductions to the normal equinox of 1875.0 were performed with Struve's constant of precession while later on New-comb's constant was generally used. The reduction to a unified system including the realisation of a true fundamental system is a real difficulty requiring ingenuity and experience.

The wealth of information on the astrometrically more important stars is gradually increasing owing to the perseverance of observers who prepare new catalogues of position; the biggest example here is the work of Yale Observatory. In most cases not only positions but also values of proper motion are published. In the routine work positions from older catalogues are introduced and the proper motions are printed in columns adjacent to the positions. With this kind of presentation the positions may be updated to any chosen epoch, but it is inadequate for improvement by the addition of further observations.

In the international undertaking AGK3, originally planned for merely furnishing (quickly) proper motions for all the stars of the AGK2, the AGK2 was revised by means of the original rectangular coordinates and, after applying systematic correc-tions, a new edition was produced in the fundamental system of the FK4. The mag-netic tapes and the manuscript (in preparation) give positions at both epochs; the printed proper motions were derived by taking the differences of positions and dividing these by the differences in epoch. The proper motions in right ascension are multiplied by the cosine of the declination in AGK3; the differences in epoch are also given. Thus it is possible not only to update the positions to other epochs but to determine the mean errors for these epochs, and to add new observations easily. In the case of more than two observations, application of Newcomb's method of the *central date* is appropriate. The addition of new observations was a special concern of the author of the *General Catalogue* (B. Boss, 1937, Vol. 1, p. 49).

In principle the zero of epoch might be chosen arbitrarily while fitting a great circle through the positions at different epochs and with different weights by least squares. If the zero of epoch is taken as the weighted mean of the epochs (the 'central date'), the resulting position for this zero epoch has maximum weight and is not dependent on the proper motions. Therefore in this case the mean error for any other epoch may be calculated directly by means of the law of propagation of errors. Furthermore, addition of new observations is possible without bias. In the present age of computers one is no longer compelled to separate the two coordinates right ascension and declination, and the fitting of observations by a great circle motion and a position angle seems to be easy enough.

3. It has often been argued that it is dangerous to use large collections of proper motions of moderate precision for the study of the kinematical behaviour of the stars in the Sun's neighbourhood. Indeed we are no longer satisfied just to repeat the pioneer work of deriving secular parallaxes from proper motions of stars of different apparent magnitude and of different galactic latitude. Nevertheless, we shall need to use the tables derived from such work for some time to come, and in the absence of other information, the reduction of plates with mean proper motion of faint stars will continue. There is a special danger in using proper motions of moderate precision for statistical purposes. The distances of the individual stars are unknown, and the restriction to directions of the proper motion recommends itself. But even here in the majority of cases the proper motion is rather small and the direction becomes meaningless, and what one is doing in averaging proper motions in small regions of the sky remains rather obscure.

In this context I would like to recall the simple concept of reduced proper motion. Assuming the absolute magnitudes of a selected class of stars to be constant, a rectification to a uniform unit of tangential velocity is possibly by reducing the proper motion μ of a star of magnitude m to a standard magnitude m_0 by

$$\log \mu' = \log \mu + 0.2 (m - m_0).$$

Thus the concept of a mean proper motion of a selected class of stars, supposed to be within a restricted range of distances in a selected region in the sky, may be transformed to the average behaviour of a whole cloud of stars surrounding the Sun, and the selected region, by inclusion of stars in a large range of apparent magnitudes, may be replaced by a selected plane through the Sun. The reduced proper motion then may be visualized as the tangential components of the space velocities; further reduction into units of $\mathrm{km\ s^{-1}}$ or $\mathrm{AU\ yr^{-1}}$ is then possible by using a common factor.

At the computing centre in Hamburg, I have the AGK3 reduced to galactic coordinates, in background storage. The standard values of Oort's rotation constants and the values recommended by Fricke for correction of precessional constants have been applied. In due course with a proper weighting system, to diminish the influence of proper motions which are small compared with their mean errors, I hope to investigate the solar apex and axes of the velocity ellipsoid for various spectral classes.

I perhaps may be permitted to offer a little warning; with the invention and installation of modern astrographs and with scanning and measuring machines operating automatically, it becomes possible to handle millions of positions and proper motions using methods of reduction which promise excellent returns in relation to effort. Before engaging in a large programme, considerable thought must be given to its objective and to what additional data such as magnitudes, colours and spectral types, may be necessary. Some positional catalogues already contain such data.

4. Special objects for proper motion study may be anything astronomers are interested in; such as classes of variable stars which may be used for studying the distance scale of the Universe, calibration of luminosity classes, moving clusters, and even globular clusters. The latter, on account of their large distances, are well suited for studying the behaviour of the surrounding field stars. The brighter stars normally have a history from which proper motions can be determined by the methods which I have already described. With powerful instruments of longer focal length, photographic methods come into the picture. Pairs of plates taken with the same instrument at different epochs give the coordinate differences in units of – say – mm. Using least squares, with linear formulae in the coordinates, the relative scale and orientation of the plates can be determined from reference stars near the edge of the field. The residuals are then converted into relative proper motion by applying a mean scale value ($''$/mm) and the time interval. The relative proper motions can then be reduced to an absolute system by assuming the average motion of a number of faint reference stars. Alternatively, if stars with known fundamental proper motions are included in the measures, the zero-point of proper motions may be referred to the fundamental system, provided that no serious magnitude error is thereby introduced. It should be stressed that, in any case, the scale value must be known a priori, and not derived from the adopted proper motions.

In cluster work it is a good policy to use the member stars themselves as a reference frame in order to achieve a good plate solution. Reference stars in the field may once more serve to reduce the proper motions to the fundamental system.

In conclusion I would like to stress the general importance of proper motion work in Kapteyn's selected areas; here we have excellent samples of data that are particularly suitable for statistical studies into greater depths of the galactic system.

DISCUSSION

Luyten: If it is appropriate to mention it here – since Dr Dieckvoss has spoken of proper motions for special objects, I should like to mention that since I have both the original Palomar Survey plates as well as repeats of all the red plates, I am in a position to determine the proper motion for *any* object brighter than 21 m_{pg} north of $-33°$ (1975).

Eichhorn: Rigorous formulae have recently been published in the Astronomische Nachrichten which give right ascensions and declinations at any epoch as rigorous functions of initial position, proper motion, and ratio of radial velocity to distance, as well as the proper motion components and the radial velocity over distance ratio at any epoch as functions of the above-mentioned values at a zero epoch. It should be pointed out that the second order formulae in most, if not all, text books are wrong since they take no account of the radial velocity to distance ratio. Since completely rigorous formulae are now available, I believe that they should be used, lest investigators come to regard the incomplete second order formulae as completely satisfactory for all stars and all epoch differences.

Dieckvoss: At present time this only applies to two stars, Barnard's, and Groombridge 1830.

Bok: Secular parallaxes for stars at large are of little use – BUT – secular parallaxes for special groups, K giants or dwarfs, for example can be very useful.

Proper motions are urgently needed for RR Lyrae variables. Dr van Herk has work under way for 400 of these stars and this project – combined with the McCormick work – should be very important for calibration purposes of absolute magnitudes. Radial velocity work (with image tubes!) on RR Lyrae stars with known p.m.'s promises to yield important results relating to the scale of our Galaxy and related problems.

Fricke: My comment consists of two parts, the first concerning Dr Eichhorn's remark on the variation of proper motions due to the foreshortening effect. This effect is very small, and the formulae given in the textbooks have been sufficient for most of the practical purposes. I think that people who derive highly accurate proper motions will not use textbooks but more rigorous formulae whenever this is necessary. Second, I want to comment on the effects of inaccurate proper motions on kinematical results. For stars nearer than about 100 pc from the Sun the effects of precessional corrections and systematic differences between modern catalogues are small in the transverse motions to be derived. Their importance grows with increasing distance from the Sun, and they all have to be taken into account in kinematics of stars farther away than 100 pc. For distances larger than about 1500 pc, even the best available proper motions (with m.e. of $\pm 0\overset{''}{.}20$ per century) fail to contribute to significant results. Furthermore, I want to say that the results of a good number of older determinations of secular parallaxes may be wrong due to the fact that the material on which they were based does not fulfil the conditions which underly the equations of determination.

Bok: Dr Fricke's condemnation of secular parallaxes is obviously justified for research at low galactic latitudes. I take the view that secular parallaxes at high *and* at intermediate galactic latitudes can serve a very useful purpose in checking on any conclusions relating to density and velocity distributions of special groups of stars perpendicular to the galactic plane. Secular parallaxes should be most useful in providing us with checks on the tilts of equidensity surfaces at various points above and below.

Fricke: I should have certainly specified in more detail my criticism of secular parallax determinations. One needs to know photometric data and MK spectral classes (except that one is dealing with a class of variables) in order to arrive at a significant luminosity determination. Furthermore, it is not possible to average over a large distance interval when the distribution of stars is not Gaussian about a 'mean' distance, or when the distribution deviates appreciably from a uniform one.

PROPER MOTIONS WITH RESPECT TO GALAXIES

(Invited Paper)

A. N. DEUTSCH

Pulkovo Observatory, U.S.S.R.

and

A. R. KLEMOLA

Lick Observatory, Santa Cruz, Calif., U.S.A.

Abstract. At Lick the second phase of the proper motion program is in progress. In addition to generally selected stars, as was done for the first phase, so far over 30000 stars of special types of astro-physical interest and about 29000 AGK3 stars have been selected for measurement.

In accordance with the Pulkovo program, second-epoch photography with galaxies is being continued at Pulkovo, Moscow and Tashkent, and proper motions with reference to galaxies are derived.

Analyses of proper motions at Pulkovo and Lick show agreement in some instances and disagreement in others. The same applies to comparisons with fundamental catalogues. The analyses suffer to some extent from absence of proper motions in the zone of avoidance and in the southern part of the sky.

In the southern hemisphere, first-epoch photography of 164 fields with galaxies has been completed using the Maksutov double-meniscus telescope at Cerro El Roble in Chile, and a complete coverage of the sky has been started with the same telescope; this work is being done jointly by the Soviet and Chilean astronomers. On the Yale-Columbia southern program, the first-epoch photography is nearly completed with the double astrograph at Leoncito in Argentina. There are plans at Lohrmann Institute, Dresden, to take photographs with the 2-m Schmidt telescope at Tautenburg, thus providing first-epoch plates for proper motions with reference to galaxies.

1. Introductory Note

One of the authors (A.N.D.) was not able to attend this symposium and his contribution communicated by correspondence has been incorporated in this report by the other author (A.R.K.).

2. Summary of Proper Motion Programs

The nature and present status of the various programs for the measurement of stellar proper motions with respect to galaxies is given in Table I. These comments depend in part on a review article by Vasilevskis (1973).

In the table are given the main elements of the Lick and Yale-Columbia programs based on 51-cm aperture astrographs, as well as the original program of the Pulkovo Observatory using various 33-cm Normal Astrographs. A recent addition is the 70-cm double meniscus Maksutov telescope in Chile. And finally there is the declared intention of H.-U. Sandig at Dresden to use the 2-m Universal Reflecting Telescope of the Karl-Schwarzschild Observatory in Tautenburg for the measurement of proper motions in selected fields. Descriptions of each of the programs is given in the following sections except for the Yale-Columbia program which is described by A. Wesselink elsewhere at this symposium.

Gliese, Murray, and Tucker, 'New Problems in Astrometry', 193–200. All Rights Reserved.

A. N. DEUTSCH AND A. R. KLEMOLA

TABLE I

Programs for proper motions

Program	Lick	Yale-Columbia	Pulkovo I	Pulkovo II	Lohrmann
Location	Mt. Hamilton	Leoncito	(various)	C. El Roble	Tautenburg
Latitude	$+37°$	$-32°$	(various)	$-32°$	$+51°$
Telescope[a]	51-cm DA	51-cm DA	33-cm NA	70-cm DMM	134-cm
$f/$	7.3	7.3	10.0	2.9	3.0
Scale (″/mm)	55.1	55.1	60.0	103.1	53.2
Field (deg)	6.3	6.3	2.0	5.	3.4
Mag. Limit	19B, 18V	19B, 18V	14	19	20B
Sky Region	$+90°$ to $-23°$	$-90°$ to $-22°$	galaxies	S. sky/galaxies	N. sky zones
No. Fields	1246	598	300:	$-/164$	1400

[a] Telescopes are denoted DA = Double Astrograph, NA = Normal Astrograph, and DMM = Double-Meniscus Maksutov. Aperture of the Schmidt corrector given for the 2-m Tautenburg telescope.

2.1. LICK PROGRAM

The distribution of fields and status of observations is shown in Figure 1. The total number of fields is 1246 and 1390 to southern declination limits of $-23°$ and $-33°$, respectively. Although the original program as proposed by W. H. Wright extends to a limit of $-23°$, an examination of supplementary plates taken by Shane and Wirtanen reveals that plates with centers at $-25°$ are of acceptable quality for measurement as well as the central areas of plates with centers at $-30°$. It is estimated that the number of fields with a sufficient number of galaxies to obtain plate constants amounts to 926 and 1031 to $-23°$ and $-33°$, respectively. Therefore the fraction of fields lying outside the zone of avoidance is 74%. As of 1973 June the total number of fields with acceptable second-epoch plates came to 336. These are located mostly away from the Milky Way, as seen from Figure 1.

One of the main goals of the program is the determination of corrections to the motions in the FK4 system and the constants of precession. Since the stars of the FK4 are too bright for measurement on blue astrograph plates, it is necessary to select fainter stars from the AGK3. It should be pointed out that the yellow plates may be used at a future third epoch to measure motions of the fainter FK4 stars, say as bright as magnitude 5.0, since the growth of image size with magnitude is much smaller than for the blue plates. The selection of stars in three ranges of magnitude was made from a magnetic tape containing the AGK3. The density of selected stars on the sky from declination $+90°$ to $-2°.5$ is shown in Table II. The number of stars selected is 28 821 from a total of 183 572 in the AGK3, or 15.8%. If an estimated 70% survive to the final stage of reduction, then the final number will be near 20 000, or 11.0% of the AGK3. Since this figure includes the sky within the zone of avoidance, the number lying outside and thus measured directly with respect to galaxies should be about 74% of this figure, or 15 000 stars.

Catalogue stars for the magnitude range 7 to 12 for the sky southward from dec-

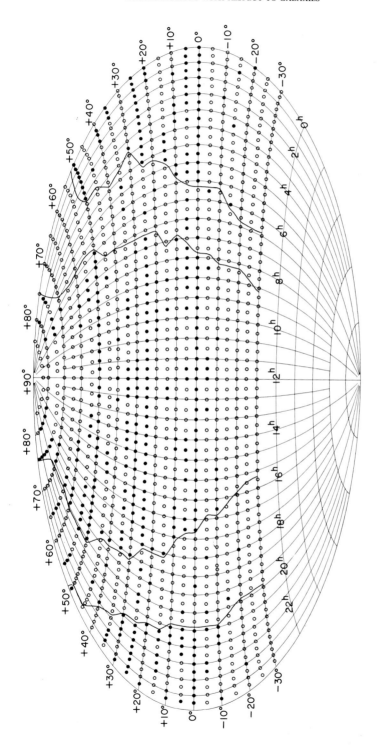

Fig. 1. Status of Lick proper motion program. Filled circles represent fields with acceptable second-epoch blue and yellow photographs.

TABLE II

Number of stars per square degree selected from AGK3

AG	Group	Bright	Medium	Faint
Zone	Mag.	7.0–8.5	8.6–10.2	10.3–12.0
+89° to +86°		0.33	1.0	1.0
+85° to −1°		0.33	0.5	0.5
−2°		0.66	1.0	1.0

lination −2°5 will be taken from the *Smithsonian Astrophysical Observatory Catalogue* with a density the same as for the northern sky for the three magnitude groups.

To permit the transformation of iris photometer readings into magnitudes and color indices in the *B, V* system, stars are selected from the photometric catalogue of Blanco *et al.* (1968). Most of these stars are brighter than mag. 13, so that only an approximate reduction will be possible for the faint program stars in most fields.

One of the important expected contributions from the program is the measurement of proper motions for various classes of stars chosen from the literature on the basis of probable astrophysical or kinematical interest. At the meeting of the IAU at Brighton in 1970 it was reported that as many of the RR Lyr stars as possible would be measured. Since then the scope of the program has been enlarged greatly. Now a selection of all available members of many classes of stars has been undertaken. The details of the selection are shown in Tables III and IV. In all there are about 30 000 stars of a special character. The inclusion of new stars continues as new publications are received. Duplicate entries for the same star will be sorted out as the stars are viewed in the survey machine.

In Table III an asterisk denotes an incomplete selection of a particular class of variable star. It is not intended to include objects for which the types are unknown or are constant in light. Supernovae are excluded. At least a partial selection of eclipsing systems will be made. The total selection north of declination −33° amounts to 7424 variables of the total 20 448 given in the *General Catalogue of Variable Stars* (Kukarkin *et al.*, 1969).

In Table IV are given other classes of stars, some of which are found in the listing of variable stars. In addition QSO's and Zwicky's compact galaxies are included in the program. These will serve as additional reference objects. Finally fainter stars appearing in the lists of the International Polar Motion Service for observation with PZT's have been selected.

By measuring motions for as many as possible of a given class of star, it will be possible to attempt finer division into subgroups for analysis of smaller variations of mean absolute magnitude. To permit maximum usefulness of the proper motion data, it would be desirable that radial velocity and photometric data be obtained for as many of these stars as possible. Later when the Yale-Columbia southern pro-

TABLE III

Selection of variable stars
from *General Catalogue of Variable Stars* (1968)

Type	GCVS (1968)	North −33°
Cepheid	706	441
Irregular	1687	1458
Mira	4566	826[a]
Semi-regular	2221	1733
RR Lyrae	4433	2426
RV Tauri	104	79
β Cephei	23	17
δ Scuti	17	17
α^2 CVn	28	27
All Pulsating	13782	7024
Nova/Nova-like	203	153
R Cor Bor	32	15
Irregular	1109	0[a]
UV Ceti	28	24
U Gem	215	162
Z Cam	20	18
Supernova	7	0
All Eruptive	1618	372
Eclipsing	4062	0[a]
Unique	35	28
Unstudied	803	0
Constant	148	0
All Other	5048	28

[a] Very incomplete selection.

gram is completed, there will be on hand motions for variable and non-variable stars for the whole sky. This full sample should then permit a better determination of stellar motions and catalogue corrections than is possible today.

As in the pilot program, faint anonymous stars are selected for measurement. Experience with the pilot program showed that a wider range of magnitude was desirable. Consequently two stars per square degree are selected: one of mag. 14–15 and another of mag. 15–17 for a total of 72 per field of 6° × 6°.

The selection of galaxies remains the same as for the pilot program, namely, at least one per square degree where possible (72 per plate and up to 100 in rich fields).

2.2. PULKOVO PROGRAMS

The Pulkovo proper motion programs may be discussed in three parts. First, there is the original program using various Normal Astrographs for which results of mea-

TABLE IV

Selection of other classes of objects

Class	Number	Class	Number
Ap, Am	394	Metal-deficient	300
Barium	150	Miscellaneous	1215
Carbon (R, N)	1629	O–B	94[a]
Compact galaxies	1934	Planetary nebulae	485
Common proper motion	892	QSO	545
Faint blue objects	6226	Red (giants, dwarfs)	6850
Helium	10	Subdwarfs	416
Horizontal-branch	53	White dwarfs	900
Infra-red	285	Wolf-Rayet	66
IPMS (PZT)	96[a]	X-ray	2
		Sum	22542

[a] Very incomplete selection.

surements and analysis have been published for 85 fields. Second, there is the current program which is the continuation of the second-epoch photography with the Normal Astrographs. Deutsch in his report to Commission 24 reports that plates taken at Pulkovo from 1949 to 1958 serve for the first epoch and are repeated once the epoch difference reaches 20 years. Stars down to mag. 14 are measured. At Moscow and Tashkent the new plates are compared with those taken before 1941. The third part of the Pulkovo programs consists of the application of the 70-cm double-meniscus Maksutov telescope at Cerro El Roble for a southern program in two parts: (1) the recently completed first-epoch photography for 164 fields with galaxies and (2) the recently started full coverage of the southern sky with blue and yellow plates. Plates are taken with long (30 min) and short (1–3 min) exposures and will serve to determine absolute proper motions of stars down to magnitude 19.

At Pulkovo (comments by A.N.D.) preliminary results of new measurements with the semi-automatic measuring machine ASCORECORD confirm the earlier conclusion that

$$\text{p.e. (galaxy)} = 1.5 \text{ p.e. (reference stars, } \bar{m}_{pg} = 13.8)$$

and

$$\text{p.e. (galaxy)} = \text{p.e. (AGK3 star).}$$

The measured 'proper motion of a galaxy' with respect to faint reference stars for one plate pair is $\pm 0\rlap{.}''008$ to $0\rlap{.}''010 \text{ yr}^{-1}$. It is expected that future measurements of the second-epoch Normal Astrograph plates will yield a p.e. for a galaxy or AGK3 star of $\pm 0\rlap{.}''002$ to $0\rlap{.}''003 \text{ yr}^{-1}$ from about 12 stars in a circle 50' radius from 3 pairs of plates with an average of 3–4 galaxies per field.

The belief is further emphasized that the best method of measurement is through the use of a blink comparator. It should be faster and more precise since the measurer sees both plates simultaneously. However, such a device as envisioned here does not now exist.

The analysis of proper motions yields a solar apex that differs from the standard apex for both Pulkovo and Lick with the former being even further off than the latter. Proper motions obtained from early studies of Selected Areas showed the same effect but to a lesser degree. The Pulkovo mean secular parallaxes appear twice as large as in previous investigations, particularly at high galactic latitudes. The outstanding feature of the study of the differences in proper motions between Pulkovo and the AGK3 is that the run of differences in proper motions in declination as a function of declination reaches $0''\!.007$ yr^{-1} for declinations south of $+40°$. This is not observed for the Lick motions.

2.3. LOHRMANN PROGRAM

Plans are underway by H.-U. Sandig of the Lohrmann Institute in Dresden, GDR, to use the 2-m Universal Reflecting Telescope of the Karl-Schwarzschild Observatory used in the Schmidt-system mode in a program of proper motions with respect to galaxies. The purpose of the program is the determination of errors of meridian circle catalogues after repetition of the plates in about 20 years.

Fields selected for the program will overlap center-to-center. The fields to be observed lie along four parallels of declination: $0°$, $+26°$, $+52°$, and $+75°$. In addition, other fields to be observed lie along hour circles in the northern hemisphere at every two hours of right ascension. In all there will be about 1400 plates of 10^m exposure each reaching mag. 20. Deutsch suggests that the fields of galaxies of the Pulkovo program be observed also. This would involve two plates for each of 140 fields.

3. Some Outstanding Problems

Once the absolute proper motions are on hand, there still remain several outstanding problems that hinder a best possible determination of solar motion, galactic rotation, and mean absolute magnitudes.

There is a distinct need for adequate magnitude and color standards for the fainter stars, say beyond mag. 13. As the determination of these photometric quantities falls outside the scope of the observations being carried out at Lick, it is hoped that such work will be carried out at other observatories.

Another problem is the need for adequate correction for interstellar absorption and reddening. Since stars as faint as mag. 18 are observed, corrections for the more luminous stars will be needed to great distances from the Sun. Representation of the absorption and reddening by more complex methods than the commonly used cosecant or exponential formulae would seem more realistic. One approach could be the use of the detailed analysis by Fitzgerald (1968) based on stars from the photometric catalogue of Blanco *et al.* (1968) which is modified for rapid computer reductions.

The bridging of the zone of avoidance may be approached from several ways. As seen earlier, about 26% of the program fields at Lick have an insufficient number or poor distribution of galaxies to permit the determination of plate constants. However, a portion of these Milky Way fields do have several galaxies, so that zero-point corrections to arrive at absolute motions in the manner of Pulkovo may be obtainable. Alternatively, the AGK3 stars, with motions corrected by extrapolated values found at higher latitudes, may be employed in the zone of avoidance. The various methods may be tested.

The programs now in progress should eventually provide a much stronger observational basis for the system of proper motions as well as the conclusions concerning solar motion, mean parallaxes, and galactic dynamics.

References

Blanco, V. M., Demers, S., Douglass, G. G., and Fitzgerald, M. P.: 1968, *Publ. U.S. Naval Obs., 2nd Ser.* **21**.

Fitzgerald, M. P.: 1968, *Astron. J.* **73**, 983.

Kukarkin, B. V., Kholopov, P. N., Efremov, Yu. N., Kukarkina, N. P., Kurochkin, N. E., Medvedeva, G. I., Perova, N. B., Fedorovich, V. P., and Frolov, M. S.: 1969, *General Catalogue of Variable Stars* (GCVS 1968).

Vasilevskis, S.: 1973, *Vistas in Astronomy* **15**, 145.

DISCUSSION

Bok: When may we expect to have the proper motion for the RR Lyrae variables outside the zone of avoidance, say 2000 stars? What will be your Lick magnitude limit?

Klemola: The main work is concerned with that part of the sky outside the zone of avoidance. This is about 74% of the full programme. The results should be on hand in about 4 or 5 years. A useful magnitude limit is about 17.5.

Murray: Could you consider including some faint red stars, say 16–17 mag., in your survey? These would have to be selected photometrically; perhaps Prof. Luyten has plate material from which extreme red faint objects could be detected.

Klemola: Yes, faint red stars could be included. What star lists would you suggest?

Murray: This would need some careful planning.

Brosche: What are the measurement principles for galaxies?

Klemola: The scanner of the measuring machine performs centring on some sort of density centre.

Luyten: Since I determine relative proper motions on the Palomar plates I am very much interested in obtaining data which would give us the correction from relative to absolute motion for various magnitudes all over the sky. How soon can we expect this?

Klemola: Present efforts are concerned with zones around declinations 0°, +25°, and +50°. These partial results should be available in a year or two.

REPORT ON THE YALE-COLUMBIA SOUTHERN PROPER MOTION PROGRAM

A. J. WESSELINK

Yale University Observatory, U.S.A.

This program is best introduced by saying that it constitutes the southern counter-part of the well known Lick proper motion program. In fact, it has been designed to be almost identical to that so that when the work on the materials from both hemispheres has been completed, a homogeneous set of data will be available for the entire celestial sphere.

The Lick program has been described on several occasions by Vasilevskis and Klemola (Vasilevskis and Klemola, 1970; Deutsch and Klemola, 1973, 1974). Since the Lick astronomers are about twenty years ahead of us, they are solving many problems of our program even before we become aware of them. We are indebted for their leadership in many ways.

The details of the program and the progress made at the time of writing this Report are given in the Table I.

The Southern Proper motion program was initiated by D. Brouwer and J. Schilt as a joint enterprise of the Yale and Columbia Universities. It is financially sup-ported by the National Science Foundation. The telescope is a double astrograph; the two refractors being designed respectively for photography in the blue and in the yellow. It is located at El Leoncito, a remote place in the Western Andes in the province of San Juan in Argentina, at an altitude of 8000 ft. Observing of the first epoch plates was begun vigorously only in 1966 with Klemola as the first observer. A. G. Samuels took over in 1967 and continued until 1971 when J. Gibson succeeded; he is still continuing.

The objective gratings and the two exposures of different duration both serve to give proper motions for stars between 6^m and 18^m essentially free of magnitude error.

We mentioned in the introduction that for the solution of many of our problems we receive guidance from the Lick astronomers, for which we are duly grateful. How-ever, there are difficulties which are entirely our own. When it became known that the NSF support would be only moral as from January 1, 1974 some changes had to be made in the observing policy to save the first epoch to best advantage.

Originally we pursued observing the same field until acceptable plates had been secured in the blue as well as in the yellow. As from March 1971 the observations are continued in both colours as before. However, a field is considered satisfactorily observed whenever a good blue plate has been secured, regardless of the quality of the yellow photograph. With the new policy we may reasonably hope that the blue plates are all of acceptable quality and have been taken for 99% of Program I at the time financial support is ending.

Gliese, Murray, and Tucker, 'New Problems in Astrometry', 201–203. All Rights Reserved.

TABLE I

Details of telescope and program

	Blue	Yellow
Aperture of refractors:	20 in.	20 in.
Mean effective wavelength:	4300 Å	5550 Å
Mean scale 1 mm:	55″036	55″136
Size of plates:	17 × 17 in.	
Size of plates:	6°3 × 6°3	
Designer of optics:	J. Baker	
Manufacturer of optics:	Perkin Elmer	
Designer of mechanical parts:	B. Hooghoudt	
Manufacturer of mechanical parts:	Rademakers	
Distance of adjacent declination zones:	Five degrees	
Distances between centers of adjacent fields within same zone do not exceed five degrees.		
Plate overlap:	At least one degree	
Difference in magnitude between central image and first order images due to objective grating:	3.5 mag.	
Exposure times of two exposures on same plate:	Two hours and two minutes	
Limiting magnitude long blue exposure:	19^m	
Limiting magnitude long yellow exposure:	18^m	
Expected accuracy of annual proper motions with twenty years between epochs:	$\pm 0″005$	
Program I: zone $-25°$ and further south including $-90°$.		
Program II: zone $-20°$ and further north including the zone of $0°$ declination.		
Total number of fields Program I:	598	
Total number of fields Program II:	360	
	958	
Total number of fields satisfactorily photographed (status in July 1973) at the first epoch:	Program I: 560	
	Program II: 63	

We are steadily working on Program II, the overlap with the more northern fields already observed by the Lick observatory. For Program II we follow the same policy with regard to acceptance of blue and yellow negatives as for Program I.

It is certain, that Program II which has second priority after Program I, will not be finished at the time of closure of the southern observatory. However, enough plates of this program have been secured on the first epoch that a worthwhile comparison with Lick results may be anticipated in due course (Deutsch and Klemola, 1973, 1974).

There has been some discussion in the past about the number of years that Yale-Columbia shall allow to lapse before beginning the observing for the second epoch. Vasilevskis and Klemola maintain that an interval of 10 yr with the southern telescope will give an accuracy comparable to 20 yr with the Lick instrument (Vasilevskis and Klemola, 1970). It is true that a number of problems can be solved satisfactorily with an interval as short as ten years between epochs.

On the other hand there are problems which require a higher accuracy than is obtainable with our telescope in ten years. It is in these cases that a longer time base will only be good enough. It is for these problems that Prof. Blaauw recommended an interval as long as 20–30 yr. There are factors, other than astronomical, which may decide what is to be done in the future and when. Whatever will happen cannot now be predicted and depends on the interest, patience and financial resources of future generations. Few things improve by postponement of action: Proper motions do.

References

Deutsch, A. N. and Klemola, A. R.: 1973, *Contrib. Lick Obs.*, No. 399.
Deutsch, A. N. and Klemola, A. R.: 1974, this volume, p. 193.
Vasilevskis, S. and Klemola, A. R.: 1970, in W. J. Luyten (ed.), 'Proper Motions', *IAU Colloq.* **7**, 167.

DISCUSSION

Bok: We should speak out strongly in support of the effective rounding out of the Yale-Columbia programme. We should stress that it is in the interest of fundamental research that the first phase of the programme as originally proposed should be *completed in full* and that the second phase – the taking and measuring of the second epoch plates for determinations of proper motions – should be undertaken at a time set by purely scientific considerations.

COMPARISON OF SECULAR PARALLAXES DETERMINED FROM PROPER MOTIONS OF STARS RELATIVE TO GALAXIES AT THE PULKOVO AND LICK OBSERVATORIES

N. V. FATCHIKHIN

Pulkovo Observatory, U.S.S.R.

The secular parallaxes, determined by Klemola and Vasilevskis from proper motions of stars relative to galaxies (Klemola and Vasilevskis, 1971), were interpolated in latitude for each area of the Pulkovo Catalogue (Fatchikhin, 1974). Then the mean values were grouped in three zones of galactic latitude (b) and the data obtained at the Lick Observatory (L) were compared with our data (P) (Fatchikhin, 1972, 1973). The results are given in the Table I.

TABLE I

In 0″.001

m	9.5	10.5	11.5	12.5	13.5	14.5	15.5
(1) Galactic latitude from $\pm 11°$ to $\pm 25°$ ($\bar{b}=18°$)							
L	20.7	15.5	12.6	10.9	10.0	9.4	9.0
P	18.6	16.2	12.5	10.3	9.5	6.6	3.8
$L-P$	$+2.1$	-0.7	$+0.1$	$+0.6$	$+0.5$	$+2.8$	$+5.2$
(2) Galactic latitude from $\pm 25°$ to $\pm 50°$ ($\bar{b}=39°$)							
L	24.6	20.1	16.5	14.3	12.8	12.9	11.2
P	20.5	17.6	13.8	10.5	8.7	8.2	7.5
$L-P$	$+4.1$	$+2.5$	$+2.7$	$+3.8$	$+4.2$	$+3.7$	$+3.7$
(3) Galactic latitude from $\pm 50°$ to $+90°$ ($\bar{b}+67°$)							
L	28.9	24.9	21.8	19.1	17.1	15.4	14.2
P	26.2	25.7	23.8	19.2	14.2	10.7	9.1
$L-P$	$+2.7$	-0.8	-2.0	-0.1	$+2.9$	$+4.7$	$+5.1$

As seen from the table the greatest deviation in the parallaxes is in the middle zone ($\bar{b}=38°$): $\overline{L-P}=+0″.0035$.

Previously we compared our results with those obtained at the Groningen, McCormick, Radcliffe Observatories and also with the Pulkovo results in Kapteyn's areas (Fatchikhin, 1972, 1973). On the base of this comparison we conclude that secular parallaxes of stars in high galactic latitudes ($\bar{b}=67°$) obtained recently at the Pulkovo and the Lick Observatories appeared to be larger than the data of the above mentioned Observatories. The mean values of secular parallaxes were calculated from our data (Fatchikhin, 1974) in the 10° zones of galactic latitude and the 90° longitude (Fatchikhin, 1972, 1973). The wave-like variations of these parallaxes have been detected, which could be explained by rotation of a local star system.

Gliese, Murray, and Tucker, 'New Problems in Astrometry', 205–206. All Rights Reserved.
Copyright © 1974 by the IAU.

References

Fatchikhin, N. V.: 1974, 'Catalogue of 14600 Stars', *Trudy Pulkovo Obs.* **81**, in press.
Fatchikhin, N. V.: 1972, *Astron. Circ. Acad. Sci. U.S.S.R.*, No. 668.
Fatchikhin, N. V.: 1973, *Astron. Zh.* **50**, 2.
Klemola, A. R. and Vasilevskis, S.: 1971, *Publ. Lick Obs.* **22**, 111.

THE ERRORS OF ABSOLUTE PHOTOGRAPHIC PROPER
MOTIONS OF STARS RELATIVE TO GALAXIES

A. A. KISELEV

Pulkovo Observatory, U.S.S.R.

In Photographic Astrometry the absolute proper motions of stars are determined relative to galaxies by the direct (I) or indirect (II) methods. In the direct method the plate constants are determined directly from galaxies, used as the reference stars with a Zero proper motion. In the indirect method the proper motions are first found relative to the totality of reference stars S_r ($r = 1, 2, \ldots r_0$), then the obtained relative proper motions are reduced to absolute values by means of galaxies G_g ($g = 1, 2, \ldots g_0$), the correction, identical for all the stars, being derived by formula (I).

$$\Delta \mu_s = \mu_s - \mu_s' = -\frac{1}{g_0} \sum_{g=1}^{g_0} \mu_g' \tag{1}$$

In the classical case (reduction by the linear formulas) method I provides a geometrically strict solution. The mean error of the absolute proper motion of a star S_s as derived by method I is determined by formula (2)

$$E_{\mu s (I)}^2 = \frac{E_{\Delta x S}^2}{\tau^2} + \frac{E_{\Delta x G}^2}{\tau^2} \sum_{g=1}^{g_0} D_{gs}^2, \tag{2}$$

where $E_{\Delta x S}$ and $E_{\Delta x G}$ are the mean errors of measurements of one difference of coordinates for the stars and galaxies on the plates of the first and second epochs, D_{gs} – conditional weights (Schlesinger's dependences) of the galaxies relative to the stars S_s, and τ – the difference of epochs.

Method II does not provide the strict geometrical solution of the problem, however, with respect to accidental errors, it can result in more precise solution, as well as in less precise one. The mean error of the absolute proper motion of a star, as derived by method II, is determined by formula (3)

$$E_{\mu s (II)}^2 = \frac{E_{\Delta x S}^2}{\tau^2} + \frac{E_{\Delta x G}^2}{\tau^2} \frac{1}{g_0} + \left(\sigma_R^2 + \frac{E_{\Delta x R}^2}{\tau^2} \right) \left(\sum_{r=1}^{r_0} D_{rs*}^2 - \frac{1}{r_0} \right). \tag{3}$$

Here $E_{\Delta x R}$ – a mean error of measurements of the coordinates difference for the reference stars, σ_R – a dispersion of relative proper motions of the reference stars, used in determination of the initial relative proper motions, D_{rs*} – conditional weights of reference stars relative to a fictitious object placed at the point with the coordinates $x_s^* y_s^*$, complying with the condition (4)

$$x_s^* - \bar{x}_R = x_s - \bar{x}_G; \quad y_s^* - \bar{y}_R = y_s - \bar{y}_G. \tag{4}$$

Gliese, Murray, and Tucker, 'New Problems in Astrometry', 207–208. All Rights Reserved.
Copyright © 1974 by the IAU.

$x_s y_s$ – are the coordinates of the star under determination $\bar{x}_R \bar{y}_R$ and $\bar{x}_G \bar{y}_G$ – the coordinates of the gravity centers of the systems of reference stars and galaxies.

A comparison of formulas (2) and (3) leads to the following conclusions:

(1) The error $E_{\mu s}$ in method II depends considerably on the dispersion of the proper motions of reference stars. This influence became apparent when difference of epochs is large. At a small number of galaxies (of the order of 5) and the epochs difference of 30 yr Method II provides more accurate absolute proper motions than Method I in case if there will be used 40–50 reference stars with $\sigma_R \leqslant 0''.015$ per year.

(2) The error $E_{\mu s}$ in both the methods increases as the object removes from the gravity center of the galaxies, however in Method I this effect is due to the errors of measurements of galaxies, whereas in Method II to the errors of measurements of reference stars and the dispersion of their proper motions. For one object in the gravity center of the galaxies both methods give coinciding results.

(3) The errors $E_{\mu s}$ both in Method I and II can be essentially different for different stars under determination (due to the influence of configuration). This effect is to be taken into account in a comparison of photographical and meridian proper motions.

Reference

Kiselev, A. A. and Yagudin, L. I.: 1973, *Trudy 19th Astrometr. Konf.*, in press.

ON THE PROJECT OF A NEW FOURFOLD COVERAGE
OF THE NORTHERN HEMISPHERE

CHR. DE VEGT

Hamburger Sternwarte, F.R. Germany

Abstract. The fully inherent information of the AGK 2 plate material, covering the northern hemisphere down to $\delta = -2.5°$ with homogeneous epoch and limiting magnitude has not been used to establish the AGK2 catalogue. A new measurement will provide positions for all stars with at least $m_{pg} = 12$ in the FK4 system, yielding an estimated accuracy of $\sigma = 0''.14$.

In March 1973 a newly developed zone astrograph for the yellow spectral-region has been set in operation at the Hamburg observatory which would be available about 1975 for a new fourfold coverage of the northern hemisphere. A technical description of the instrument is given. Details of the fourfold plate coverage and the observing program are discussed. As a suitable reference star system the AGK3R catalogue, updated with recently derived proper motions to that epoch, is adopted. A final positional accuracy $\sigma = 0''.11$ of the new catalogue is expected. The available epoch difference of 45 years up to that date will then provide proper motions in the FK 4 system with an estimated accuracy of at least $0''.005/a$ for all stars of the northern hemisphere down to $m_{pg} = 12$.

1. Introduction

The recent installation of a newly developed zone astrograph at the Hamburg observatory has offered the possibility to consider the project of a new fourfold coverage of the northern hemisphere in the yellow spectral-region. Adopting for this catalogue a limiting magnitude of about $m_{pg} = 12$, the full magnitude range of the AGK2 plate material, which has not been utilized to establish the AGK2 catalogue, can now be exhausted. Performing a new measurement of the AGK2 plates (epoch 1930) to their limiting magnitude, a combination with the new catalogue positions will provide then proper motions in the FK4 system for all stars of the northern hemisphere down to $m_{pg} = 12$.

If the catalogue project would be realized in the next years, say 1975, it will not be necessary to establish a reference system from new transit-circle observations, because the AGK3R catalogue, updated to that epoch, can be used without appreciably influencing the inherent accuracy of the new photographic catalogue positions. With an automatic measuring machine, fully available for this catalogue work and allowing an on-line preliminary reduction of the plate measurements, the performance of the whole catalogue project should be possible within 6 years.

Together with the just finished fourfold coverage of the southern hemisphere by the Cape observatory (Clube, 1970), the whole sky will then be covered with a nearly similar photographic network, obtained in the same spectral region, and reaching at least a common limiting magnitude of $m_{vis} = 11.5$ for both hemispheres.

2. The AGK2 Plate Material

The AGK2 catalogue, planned as a repetition of the AGK1 by photographic means,

Gliese, Murray, and Tucker, 'New Problems in Astrometry', 209–215. All Rights Reserved.

covers the northern hemisphere down to $-2°.5$ Decl. The plate material, which will be considered here, had been obtained at the observatories of Bonn (720 plates, Decl. between $+20°$ and $-2°$) and Bergedorf (1219 plates, Decl. between $+90°$ and $+20°$) with two similar 'AG-type' zone-astrographs as quoted at the beginning of the following section. The two zones have a small common overlapping region, extending to $2°.6$ in Decl. All plates were taken in a time interval of 3 years, centered around 1930.0. The adopted plate overlap pattern was a center-edge pattern, so that each star appears on at least two plates (see Figure 3, upper part), the plate size is $5° \times 5°$.

According to the guidelines of the AG-zone commission (Bauschinger, 1927) each plate contains 2 exposures with a duration of 10 min and 3 min, respectively, the short exposure was taken mainly for identification purposes. For obtaining the catalogue positions only the 10 min image was measured, and in accordance with the limiting magnitude of the (visual) AGK1 catalogue the majority of the measured stars have magnitudes up to about $m_{pg} = 9.5$.

The great value of this plate material for future astrometric work arises from the fact that the photographic plates in reality go far beyond this limiting magnitude. With the 10 min exposure all stars with at least $m_{pg} = 12$ were recorded, whereas the 3 min exposure includes already all AGK2 catalogue stars and a limiting magnitude of about $m_{pg} = 11$ has been reached.

The AGK2 plate material therefore contains almost all stars ($\geq 90\%$) of the Astrographic Catalogue but in contrast to the latter, the whole northern hemisphere is covered with homogeneous limiting magnitude, and all plates have been taken at nearly the same epoch. The complete plate material is still immediately available at Hamburg and Bonn and in excellent condition for a new measurement. Using a Galaxy-type automatic measuring machine, the new measurement of all plates to their limiting magnitude could be performed in less than 2 years.

Due to imperfections of both objectives, the Hamburg and Bonn positions are affected by systematic errors, depending on magnitude and colour and position on the plate as well (Kox, 1950). The Bonn objective has further introduced a tangential distortion, due to imperfect alignment of the components of the objective (Kohlschütter and Schorr, 1957). The colour and magnitude errors have been carefully investigated in connexion with the AGK3 work, where a new reduction of the (original) AGK2 plate measurements x; y have been made (Dieckvoss, 1960, 1962, 1971, 1973). However, these corrections have been obtained only for the magnitude range under consideration for the AGK2 and AGK3 catalogue. The proposed new measurement of the plates, including the 3 min exposure will further improve the accuracy of the star positions and strengthen the statistical significance of magnitude and colour terms over an extended range. In addition, in certain selected sky areas, cluster proper motions (de Vegt, 1973) and similar available data from the Selected Areas could be used for a closer investigation of magnitude and colour relations up to the limiting magnitude of the plates.

The new reduction of the whole plate material will be performed by blockadjustment methods (de Vegt and Ebner, 1972, 1973) using the AGK2A catalogue (transformed to the FK4) for reference star positions. The expected overall positional accuracy from the new reduction will be $\sigma = 0\overset{\prime\prime}{.}14$, for the mean plate epoch 1930.

3. The New Zone Astrograph

In 1967 W. Dieckvoss and the writer began detailed plannings for a new zone astrograph which should replace the so-called AG-astrograph ($f = 2058$ mm, F:19, field $5° \times 5°$, blue corrected four-component objective) now for more than 40 years in operation. The new instrument has been designed for about the same focal length and field size but, in opposition to the former, the objective is corrected for the yellow spectral-region, and a larger aperture has been introduced. After the necessary financial support had been provided by the Deutsche Forschungsgemeinschaft in 1968, the instrument was built by C. Zeiss, Oberkochen, and was set in operation at Bergedorf in March 1973. The main optical data of the instrument are summarized in Table I.

TABLE I

Optical data of the zone-astrograph

1. Photographic telescope		
Aperture	230 mm	
Focal length	2065 mm	
Appr. scale value	$100''$ mm^{-1}	
Focal ratio	1:9	
Field angle	$\pm 4\overset{\circ}{.}7$	
Utilized field size	$5° \times 5° \hat{=} 180 \times 180$ mm	
Plate size	$240 \times 240 \times 6$ mm	
Appr. utilized spectral range	5300–5800 [Å]	5300–6400 [Å]
Filter + emulsion	OG515 + Kodak 103aG	OG515 + $\begin{cases} \text{Ilford R 40} \\ \text{Ilford HP 3} \end{cases}$
Field distortion	$0\overset{\prime\prime}{.}02$ deg^{-3}	
Color-magnification error	$\leqslant 1\mu$ for the whole field and spectral range	
Light-concentration	90% light intensity inside a circle with 20μ diam.	
Filter size	$240 \times 240 \times 6$ mm, Schott OG 515 interchangeable	
Distance to focal plane	17 mm	
Field distortion due to filter	irregular field errors $\leqslant 0.3\mu$ in the whole field	

2. Guiding telescope	
Aperture	190 mm
Focal length	2580 mm
Pointing micrometer setting range	$50' \times 50'$
Setting accuracy	$\pm 1''$ in the whole field

Because a considerable improvement of the colour- and field-corrections as compared with the AG-objective was required, a five-element objective (Figure 1) has been designed using new types of glass for the inner lenses. The focal curve of the objective is shown in Figure 2. A maximum focus deviation of ± 0.1 mm is achieved

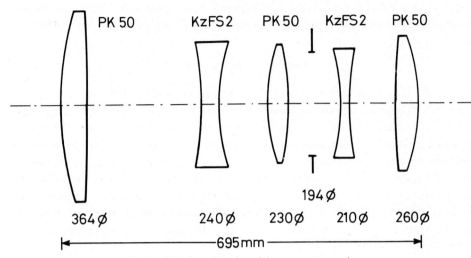

Fig. 1.   Five-lens objective of the zone-astrograph.

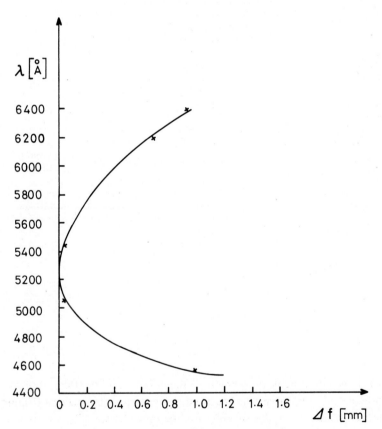

Fig. 2.   Focal curve of the zone-astrograph.

in the desired spectral range 5300–5800 Å, which is about the Rayleigh tolerance. The short wavelength cut-off is provided by an OG 515 (Schott) filter which is plane-parallel within 2". It is planned to replace this filter by an interference filter, designed for the already quoted spectral bandpass. So one would become independent of the long-wave length cut-off of the photoemulsion. At present technical difficulties have to be overcome in the performance of multi-layer filters of this large size.

The astrograph has a plane plate frame which is rigidly connected in the focal plane to the tube and defines the instrumental tangential plane. During exposure the emulsion side of the plate is pressed against the plate frame by springs. Four adjustable fiducial marks are exposed on the margin of the plate by means of a small optical imaging system. The fiducial marks give then the relation to the intersection of the optical axis with the focal plane.

The whole objective can be tilted and shifted parallel independently relative to the plate frame by a special mounting. Adjusting of the tangential point and tilt of the optical axis is performed by a small adjusting telescope, mounted perpendicularly on a plane base plate. This telescope is equipped in addition with a special auto-collimation eye piece and can be attached to the plate frame with proper orientation to the fiducial marks. Reflected images from several surfaces of the objective lenses are used to monitor the tilt of the objective. In addition, the adjusting telescope can be focussed on a small cross, placed in the middle of the last surface of the objective. With this equipment, tilt of the objective and location of the tangential point can be adjusted within an accuracy of 10".

Both coordinate-axes are provided with encoders which give a resolution of 1 s and 10" in R.A. and Decl., respectively. Hour angle and declination are displayed continuously. The polar axis of the mounting can be adjusted to any latitude between 24° and 56°. As a transportation of the whole instrument can be provided without difficulties, a temporary change to suitable observing sites for a performance of large observational programs is feasible.

4. Details of the 4-Fold Coverage

The northern hemisphere will be covered by the proposed 4-fold net down to $-2°5$ Decl. with about 4000 plates. In practice this overlap pattern would consist of two superposed nets, each similar to the AGK2 plate arrangement (see Figure 3). For one of these nets, the same plate-centers as in the case of the AGK2 will be adopted. An exposure time of 10 min will be suggested to cover the desired magnitude range and a coarse grating will be used instead of taking double exposures. Following the practice of the AGK3 work, the plates will be taken alternately with the camera west and east of its pier, in this way the influence of possible colour-magnitude errors of the objective could be eliminated. The whole observing program could be finished in about two years. With a Galaxy-type measuring machine about the same time is needed to perform the measurement of all plates.

The new zone astrograph, which is at present in its test-phase in Hamburg and will then be used for a proper motion program of open clusters (de Vegt, 1973), would be available about 1975 for a large scale observing program of this kind. As a suitable observing site the new Observatory of the Max-Planck-Institute for Astronomy in Spain is being discussed. In that case, an extension to $-30°$ Decl. in-

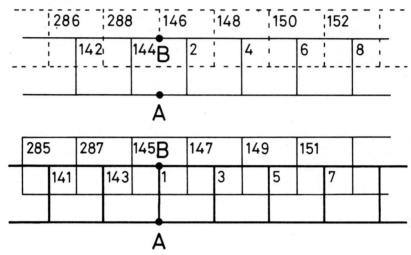

Fig. 3.   Plate arrangement for the fourfold coverage. Both nets have to be superposed on points AB.

cluding the whole zodiacal zone, which seems to be of increased importance to further investigations of the lunar orbit, could be considered. At the same time, a generous overlap with the recently finished coverage of the southern hemisphere is then obtained which allows a detailed investigation of systematic deviations of both nets.

5. Reference Star System and Photographic Accuracy

In contrast to the present situation on the southern hemisphere, no new reference star positions for the assumed epoch 1975 of the photographic coverage are available. Recently a pilot program has been started at the USNO (Corbin, 1972) to provide proper motions for the AGK3R stars from available transit-circle observations. Further on in connexion with the question of incorporating fainter stars into the proposed FK5 (Fricke, 1972), the same problem will be considered. It is therefore assumed to use the AGK3R catalogue, updated to 1975, as a suitable reference system. The final accuracy of the AGK3R stars at that epoch depends, however, strongly on the observational history of each star. Some preliminary computations, based on Corbin's material, have shown that at least an accuracy of $\sigma = 0''.18$ (m.e.) for the desired epoch should be reached. With an average number of 23 reference stars per plate and adopting $\sigma = 0''.19$ for the photographic measurements, an accuracy of $\sigma = 0''.11$ for the final catalogue positions is expected, using blockadjustment methods for the reductions.

Combining this 4-fold coverage with the newly reduced AGK2 positions, proper motions with an expected accuracy of at least $\sigma = 0''005/a$ will be obtained for all stars of the northern hemisphere down to $m_{pg} = 12$.

The extension of present catalogues, as the AGK3, to those fainter magnitudes will provide more representative samples for all investigations concerning the kinematical properties of different stellar groups. Further on, stellar positions and proper motions in this magnitude range will be of increasing importance as a second-order reference system for the evaluation of optical positions of radio sources.

References

Bauschinger, J.: 1927, *Vierteljahresschrift Astron. G.* **62**, 242.

Clube, S. V. M.: 1970, *Trans. IAU* **XIVA**, 43.

Corbin, T.: 1972, private communication.

De Vegt, Chr. and Ebner, H.: 1972, *Astron. Astrophys.* **17**, 276.

De Vegt, Chr.: 1973, to be published.

De Vegt, Chr. and Ebner, H.: 1973, this volume, p. 307.

Dieckvoss, W.: 1960, *Astron. J.* **65**, 171.

Dieckvoss, W.: 1962, *Astron. J.* **67**, 686.

Dieckvoss, W.: 1971, paper presented at the *Conf. on Photographic Astrometric Technique*, Tampa (1968), p. 161.

Dieckvoss, W.: 1973, 'Introduction to AGK 3', to be published.

Fricke, W.: 1972, in I.A.U. Colloquium No. 20, 'Meridian Astronomy', unpublished. Summarized in this volume, p. 13.

Fricke, W.: 1974, this volume, p. 23.

Kohlschütter, A. and Schorr, R.: 1957, *Zweiter Katalog der Astronomischen Gesellschaft*, Bd. 11, 27.

Kox, H.: 1950, *Astron. Nachr.* **278**, 179.

DISCUSSION

Dieckvoss stressed, that during the plate measurement automatically magnitude- and colour equivalents will be obtained (AGK2: blue, new coverage: yellow). Thus the value of the derived proper motion data will be greatly improved for all purposes of kinematical studies.

STELLAR KINEMATICS AND GALACTIC RESEARCH INVOLVING PROPER MOTIONS

(Invited Paper)

S. V. M. CLUBE

Royal Observatory, Edinburgh, U.K.

Abstract. A method of analysing proper motions as observed over all the sky is described. The kinematics of nearby stars derived from the Lick Pilot Survey are discussed, and the consequences to galactic structure examined.

The Lick Pilot Survey has provided us with a catalogue of proper motions over a wide area of the sky relative to a completely new (presumed) inertial framework. These proper motions are quite independent of the normal difficulties associated with specifying the equator and the ecliptic. Preliminary though the catalogue of data is at present, it clearly indicates that the conventional Oort-Lindblad model of stellar kinematics in the immediate solar neighbourhood is not correct (Clube, 1973). The possibility exists of course that the Lick data is vitiated by a systematic error which causes the stars to take up some spurious motion, but in spite of careful searching, there is no evidence at present for such an error (S. Vasilevskis, private communication). However, it is of some importance to note that the pattern of kinematics portrayed by the faint star proper motions bears quite remarkable resemblances to the known kinematics of intrinsically bright nearby stars – a coincidence that it would be difficult to understand were it to be attributed solely to systematic error in the proper motions. Since proper motions have in the past played a not inconsiderable role in developing our ideas regarding galactic structure and dynamics, it perhaps comes as a surprise to have it said that these recent absolute proper motions are not completely in accord with such ideas. However, the disagreement arises not so much because of any substantial difference between the Lick proper motions and those obtained previously, but more because of the use of a better kinematic model in their analysis. The precision of this model could be improved if an even greater number of absolute proper motions were available.

The great success of the Oort-Lindblad model in accounting for the main features of the kinematics of distant OB stars and cepheids (and later, neutral hydrogen) probably contributed to the decline into relative obscurity of the many papers in the 1930's which argued the lack of generality that existed in this model. Amongst these was an important paper by Milne who indicated that to a first order, the continuous stellar velocity field expressed locally by the vector **v** at a point distant **r** from the sun, could be given quite generally by:

$$\mathbf{v} = \mathbf{u} + M\mathbf{r}$$

$$M = \frac{\partial u_i}{\partial x_j} \qquad i, j = 1, 2, 3,$$

Gliese, Murray, and Tucker, 'New Problems in Astrometry', 217–220. All Rights Reserved.

where **u** is the reflex of the solar motion and u_i, x_j are velocity and distance components along (for example) the three principal galactic axes. The use of the first order Oort-Lindblad model was equivalent to setting

$$\frac{\partial u_1}{\partial x_2} = A - B, \qquad \frac{\partial u_2}{\partial x_1} = A + B$$

and all the remaining elements of M equal to zero, but it was emphasised that the real kinematic grounds for adopting the Oort-Lindblad model should have been to demonstrate statistically non-significant departures of these remaining elements from zero. This was never done, though one or two investigations gave results which implied significant departures (e.g. Mineur, 1930). In practice, the support for the Oort-Lindblad model has come mainly from other considerations and perhaps partially from apparently acceptable values for A and B. The chief reason it would seem for the failure to exploit Milne's theory fully was simply the computational problem of handling 12 unknowns. Comparative simplicity of analysis was preserved by keeping to five unknowns (**u**, A, B) though prior to the direct measurement of absolute proper motions, it has been necessary to carry two further rotational components Δn, Δk originating from errors in precession and the motion of the equinox. Now that the computing problem no longer exists, a return to Milne's theory has long been overdue.

Of course, a lack of generality exists even in Milne's model since it does not include terms in

$$x_j x_k \frac{\partial^2 u_i}{\partial x_j \, \partial x_k}$$

and other higher order terms. No analysis including such terms has yet been attempted, though there are good observational grounds for at least including

$$x_3^2 \frac{\partial^2 u_i}{\partial x_3^2}$$

in the analysis of the Lick absolute proper motions – or indeed, any others. These terms correspond to a change of secular parallax with galactic latitude regardless of sign, and although of unknown origin at present, have at least two plausible explanations, namely

(1) the Stromberg drift of high latitude faint stars with assumed larger velocity dispersion; and/or

(2) the inclusion amongst low latitude faint stars of significant numbers of very distant stars with effectively zero secular parallax.

It is unlikely however that the first order treatment of the Lick absolute proper motions will have been seriously affected by the failure to include such second order terms since the dominant effects in M are parallel to the $(x_1 x_3)$ plane and therefore virtually independent of changes in the secular parallax which are mostly in the general direction of galactic rotation.

The tensor M describing the velocity field of an effectively random selection of 9th magnitude stars in the Lick Pilot Survey is given by

$$
\begin{matrix}
0.22 & -0.35 & 0.90 \\
0.45 & (0.00) & 0.16 \\
-0.12 & -0.41 & -0.12
\end{matrix}
\quad '' \text{ per century } (\pm 0''25 \text{ approximately})
$$

The non-zero values of at least one of the elements is significant and the disagreement with the Oort-Lindblad model quite pronounced. The simplest interpretation of M would attribute it to a shear of $1''10$ about the direction $l \approx 105°$, $b \approx 20°$ with the component above the sun and mostly in the third and fourth quadrants of galactic longitude directed roughly 10° below the galactic centre, and the component below and in the first and second quadrants directed 10° above the anticentre. Since M is in fact portraying the spatial distribution of star velocity residuals after removal of the solar motion, it is pertinent to note that the direction of this shear is parallel to the principal (deviated) axis of the velocity ellipsoid of nearby stars. But even more remarkable is the fact that the 'upper' component has an almost identical motion to that of Drift II and is in that part of the sky where Gould's Belt and the Ursa Major Stream are most conspicuous, while the 'lower' component corresponds to the motion of Drift I, and is to a large extent in that part of the sky where the Hyades group predominates. The concept of stellar drifts has today fallen into disuse, but its affinity to the idea of stellar groups is very close. It has always been difficult to understand why no counterpart to Drift II has ever been discovered amongst nearby stellar groups, none of which is moving relative to the sun towards the fourth quadrant of galactic longitude. Since Gould's Belt has the motion of the Pleiades Group and the Ursa Major Stream has that of the Sirius Group, and the average of these two motions corresponds closely to that of Drift II, it seems that the quality of early data was not good enough to permit a proper separation of Drift II into these, its two principal components. It may be concluded that the 9th mag. stars have just that value for M one would expect were they to mimic the general group properties of brighter stars both in their motions and their spatial distribution. As already suggested, it would seem hardly credible that this coincidence should occur if the properties of M were in the main due to any systematic error in the Lick proper motions. It is clearly of very great importance to increase on the number of proper motions available in the present catalogue to confirm these properties of M.

If the reality of M is accepted, the consequences to galactic structure are far reaching, and these have been discussed in an earlier paper. Suffice to say that M has a rotation component about the galactic pole contrary to the direction of galactic rotation, and that this is likely to have caused serious underestimation in the past of $|B|$ especially from stars within about 300 pc of the Sun. Correcting for the effects of M from nearby stars on the more distant stars present amongst fainter magnitudes in the Lick catalogue leads to revised values for A and B and an indication that the Galaxy may be expanding violently. (It is presumed that the more distant stars do in fact move more in accord with the general predictions of the Oort-Lindblad model.) It is now very necessary to search out those observations which enable a distinction to be drawn between the conventional steady state model of the galaxy and the proposed unstable model. Perhaps this can be fairly described as 'a new problem in astrometry!'

References

Clube, S. V. M.: 1973, *Monthly Notices Roy. Astron. Soc.* **161**, 445.
Milne, E. A.: 1935, *Monthly Notices Roy. Astron. Soc.* **95**, 560.
Mineur, H.: 1930, *Monthly Notices Roy. Astron. Soc.* **90**, 789.

DISCUSSION

Edmondson: I wish to add to Dr Clube's paper that I published a paper in the *Monthly Notices* a year or two after Milne's paper in which I extended Milne's work to second order terms. The complete expansion of the second order coefficients is given in an appendix to the paper. Subsequently in another paper in the *Monthly Notices* I applied this work to the interpretation of the variation of the McCormick secular parallaxes with galactic latitude.

More important, the second order effects include terms which distort the solar motion terms in the conventional solar motion and galactic rotation solution. I have discussed this in an article in Vol. 52 of the *Handbuch der Physik* and at a Joint Discussion at the Hamburg IAU meeting.

Murray: Have you studied the M matrix for the fainter stars in the Lick Programme?

Clube: Yes. The fainter magnitude groups in the Pilot Survey are at 13th and 16th mag. photographic. After correction for the influence of nearby stars, M for the more distant stars indicates that the values of Oort's constants A and B are numerically larger, that there is a compression in the velocity field of the wider solar neighbourhood (beyond 300 pc) and that the streamlines of differential motion are not parallel to the direction of galactic rotation. Elsewhere, I have tentatively interpreted this as implying a large scale expansion of the Galaxy.

Dieckvoss: In Hamburg we have in background storage data for 180000 stars of AGK3 sorted according to spectra (HD, Vyssotsky) giving magnitude (AGK2), galactic longitude and latitude, components of proper motion in galactic co-ordinates, corrected, however, for standard Oort's A and B, and for precessional correction as recommended by Prof. Fricke. Dr Clube is invited to discuss this collection of data.

Clube: Thank you. I have in fact already examined the early type stars in the AGK3 (*Monthly Notices Roy. Astron. Soc.* **159**, 289, 1972). These also indicate that $|B|$ has a numerically larger value than the conventionally adopted one.

Fricke: Hopf was the first who suggested a more general derivation of Oort's formulae for galactic rotation including a shear and a dilation in a velocity field, in which the mean velocity of the stars per volume element is a continuous function of the distance from the Sun. Ogorodnikov (*Z. Astrophys.* **4**, 190, 1932), who has given essentially the same formulae, which were also derived independently by Milne in 1935, acknowledges that Hopf "suggested to him the general and elegant method and also contributed a great deal in working out the numerical results". Shatsova (*Publ. Astron. Obs. Leningrad* **15**, 113, 1950) has applied these formulae to the proper motions in the G.C. She came to the conclusion that a 'local stellar system' or a cloud, in which the Sun is situated, is in rapid rotation. Obviously nobody has been able to confirm her results. In analysing the FK4 proper motions in 1967, I attempted to check her results, and I found that there can only be a very small rotation in the velocity field of the stars nearer than 300 pc. The continuation of my efforts has shown that there are clear indications of deviations from the field of galactic rotation within 300 pc. The results depend on the types of the stars which are being considered. Therefore for the application of the methods developed by Ogorodnikov and Milne one needs to know MK types and photometric data so that distances can be determined and classes selected for the study of their kinematical behaviour. For the stars further away than 300 pc the Oort-Lindblad model of galactic rotation remains valid if one applies it to the mixture of stars in the disk populations. My recent results give no indication that the values of A and B which are in common use are wrong.

PROGRESS REPORT ON THE HERSTMONCEUX
PROGAMME OF MEASUREMENT OF PROPER MOTIONS
IN THE NORTHERN SELECTED AREAS

C. A. MURRAY

Royal Greenwich Observatory, Herstmonceux, England

Abstract. First results of the Herstmonceux programme of redetermination of proper motions in the Selected Areas are presented. Plate material includes the old Radcliffe plates, old plates taken at Greenwich and modern plates taken on the 26-in. refractor at Herstmonceux. The plates are being measured on GALAXY and reduced according to the plate-overlap technique.

1. Introduction

Proper motions in the northern selected areas were measured for stars down to about $m_{pg} = 15$, during the 1930's at both the Radcliffe (Oxford) (Knox-Shaw, 1934) and the Pulkova observatories (Deutsch, 1940). Extensive use of this material has been made for various statistical studies, of stellar kinematics (e.g. Hins and Blaauw, 1948) and of mean parallaxes (e.g. Binnendijk, 1943). However, both sets of data were based on rather short time bases, with a result that the optimum standard errors of individual proper motions are of the order $\pm''006$, and are larger for faint stars. As many of these faint stars will be at distances of between 500 and 1000 pc, the accidental errors are quite comparable with the likely velocity dispersions and hence will mask the true characteristics of the velocity distribution.

It was pointed out at *IAU Symp.* 1 on 'Coordination of Galactic Research' (ed. by Blaauw, 1953), that a repetition of the original plates would produce proper motions with greatly increased weight, which would make possible a much more refined study of velocity distributions of distant objects in various parts of the sky. It was also remarked that such a project would entail a large amount of labour and suggestions were then made for limiting the programme to low latitude fields.

However, with modern automatic measuring machines such as GALAXY, now available, it is possible to undertake the whole project, and indeed, the limitation of the areas measured in the original catalogues, according to galactic latitude is no longer necessary. A complete re-observation and measurement of the Radcliffe material is being undertaken at Herstmonceux, in which the full usable area of the 16 cm $\times$ 16 cm plates (scale $\sim 30''$ mm^{-1}) will be measured.

Most of the original Radcliffe plates have survived and are on loan to the Royal Greenwich Observatory through the courtesy of the Director of the University of London Observatory where the original Radcliffe 24-in. refractor is now situated. A few modern plates have been taken at London with this telescope, but in the main, the repetition has been carried out at Herstmonceux with the 26-in. refractor.

Gliese, Murray, and Tucker, 'New Problems in Astrometry', 221–225. All Rights Reserved.

In addition, plates on some areas were taken at Greenwich for photometric use, during the period 1910–20, and these have also been repeated. First indications are, however, that the original plates in this series have considerably less weight than the Radcliffe plates.

An important feature of the present programme is the measurement of B and V magnitudes as far as possible for all the stars measured for proper motion. Photometric plates for this purpose are also being obtained with the 26-in. refractor, but there is no plan at present for systematic photoelectric observations. A number of standards down to about $V = 12$ in several areas have been observed over the past few years by various RGO staff members (Epps, 1972; Penston, 1973).

As reported previously (Murray and Clube, 1970), the initial plan is to measure and study the seven areas (SA, 51, 54, 57, 71, 82, 94, 107) for which Purgathofer (1969) has published photoelectric sequences.

2. Observational Programme

Generally the Radcliffe plates on each field consist of a pair of plates between 1909 and 1919 and a pair between 1920 and 1933. Some plates however were taken on the Kapteyn plan in which first epoch plates were stored undeveloped, and re-exposed at the second epoch thus giving two epochs on the same plate. There are thus about 450 plate epochs on rather fewer plates. About 95% of these already have been re-observed at Herstmonceux.

The Greenwich photometric series contains rather less than 100 plates on 76 areas, with a heavy concentration of plates in the $+15°$ zone; the repetition of these is virtually complete.

The programme of new photometric BV plates is under way and to date, at least two plates in each colour have been taken on 24 areas.

3. Outline of Measurement and Reduction Procedures

Measurement of the plates is now in progress on the GALAXY machine at Herstmonceux.

A description of the original GALAXY at Royal Observatory Edinburgh, and its operation, have been given elsewhere (Walker, 1971; Pratt, 1971).

In the search mode, the time taken to scan a full 16 cm × 16 cm plate with the 32 μ scanning spot is about 6 h; therefore, in order to reduce the amount of searching in fields on which a number of plates have to be measured, computer programmes have been developed at RGO for producing search co-ordinates appropriate to each plate from data derived from searches on a few plates. The current procedure is to search two early plates and two modern plates, and to calculate an approximate search ephemeris for each star; in this way we hope to ensure that stars with large proper motion are not inadvertently missed, as they might well be if searches were carried out at one epoch only. The efficiency of the search mode is very dependent on plate background density, and some difficulty is being experienced with many of the old Radcliffe plates which have rather dense and variable fog levels.

All the astrometric plates are measured twice, in two opposite orientations, to eliminate magnitude equation in the measured coordinates. With this procedure we have found that the repeatability of measurement on a single plate is consistent to better than $\pm 1\ \mu$ over at least nine magnitudes.

After measurement of all plates on a field, the data are processed by a series of computer programmes in which measurements on all exposures (including multiple exposures on one plate) are transformed orthogonally to a common arbitrary coordinate system and sorted, such that all the measurements for each star are collected together, and the complete set of data is output on magnetic tape for subsequent reduction.

The reduction is carried out in four main stages:

(i) Combination of the measurements made in the two orientations.

(ii) Photometric reduction using calibration from photoelectric standards.

(iii) Transformation of all measurements to standard coordinates on a common tangent plane (including correction for refraction and aberration) through AGK3 stars, and classical solution for position and proper motion components for each star.

(iv) Final reduction of all plates on a field simultaneously by the plate overlap method.

Full details of the measurement and reduction procedures will be published in due course.

4. Preliminary Results

The computer programmes for the reduction have been developed using measurements made on 18 plates of SA54 on the GALAXY machine at Royal Observatory Edinburgh. In all, there is a total of 66 sets of measurements on 32 exposures, four of which were on modern photometric plates with single exposures and the remainder were plates ranging in epoch from 1913 to 1970, all with double exposures.

The classical solution for position and proper motion in stage (iii) of the reduction yields standard errors of unit weight for each star in each coordinate, from the residuals. In the present tests, each exposure on a double exposure plate was treated as an independent observation, and all images were given unit weight except those on the old Greenwich photometric plates which were given weight 0.1. In the X coordinates the s.e. of unit weight for stars $8 < B \leqslant 14$ is remarkably constant with a median near $\pm 0\rlap{.}''10$ and individual values ranging from $\pm 0\rlap{.}''03$ to $0\rlap{.}''26$. Since the weight of the proper motion of the majority of stars is at least 10^4, this represents a formal s.e. of a proper motion component of about $\pm 0\rlap{.}''001$. However, there is some evidence of correlation between the residuals of the two exposures on the same plate which may increase this value somewhat. In the current sample of data the Y-measurements do not appear to be as good as those in X; it is not clear at present whether this is in the plates themselves or in the measuring procedure. These plates were not measured in two opposite orientations as the presence of a magnitude equation in the GALAXY measurements was not then suspected. In both coordinates, the accuracy decreases sharply between $B = 14$ and 15 reaching a mean s.e. in X at $B = 16$ of about $\pm 0''.35$. Nevertheless this still means that the proper motions of these faint stars have formal standard errors of only about $\pm ''.003$.

High weight proper motions (wt. > 5000) have been obtained for 230 stars in the field compared with 144 stars in the Radcliffe catalogue. These include several fainter stars over the whole field as well as stars outside the area measured at Radcliffe. The proper motions are virtually complete to $B = 16$.

Only preliminary tests on the overlap reduction have so far been carried out, due to limitation of computer power, but a limited solution using plate constants and simple magnitude equation has been achieved for a small subset (10 exposures) of the total available data. In this solution the usual constraints on the relative proper motion system (i.e. no translation, shear or rotation) were applied, and also the magnitude equations were constrained so that their time weighted average was zero.

Investigations into a suitable weighting system for the final overlap solution programme are under way. It is intended that this should take account of image size (which is automatically recorded in measurement on GALAXY) as well as plate quality. Information for this is being obtained from comparison of the two exposures on each plate.

Acknowledgements

The computer programmes for this work and for general use with GALAXY have been written by a number of members of the staff of the Astrometry Department. Mr W. Nicholson, with the assistance of Mrs P. Fosbury, has developed the general purpose data processing and sorting routines for GALAXY; Mrs S. B. Tritton has written programmes for generating search data, prior to measurement and has also developed the photometric reduction programmes; the present author has been responsible for stages (i) and (iii) of the reduction outlined above, and the programme for the final plate overlap is being developed by Dr B. F. Jones.

The daily operation of the GALAXY machine is carried out by Mr E. D. Clements and Mr B. S. Carter. Mr Clements also organised the observations on the 26-in. refractor.

The efficient use of GALAXY requires considerable use of the computer and I am grateful to the staff of the Computer section of HM Nautical Almanac Office for their assistance in this and all computing work associated with GALAXY.

I am also grateful to Dr Neil Pratt of Royal Observatory Edinburgh for his help with the measurement of the SA 54 plates on the Edinburgh GALAXY, and to Mr E. W. Foster and Mr D. J. Carnochen for taking plates on the 24-in. refractor at University of London Observatory.

References

Binnendijk, L.: 1943, *Bull. Astron. Inst. Neth.* **X**, 9.
Blaauw, A. (ed.): 1955, in 'Co-ordination of Galactic Research', *IAU Symp.* **1**, 35.

Deutsch, A. N.: 1940, *Pulkovo Publ.* **II**, LV.

Hins, C. H. and Blaauw, A.: 1948, *Bull. Astron. Inst. Neth.* **X**, 365.

Epps, E. A.: 1972, *R. Obs. Bull.*, No. 176.

Knox-Shaw, H. and Scott Barrett, H. G., 1934, *The Radcliffe Catalogue of Proper Motions* (O.U.P.).

Murray, C. A. and Clube, S. V. M.: 1970, in W. J. Luyten (ed.), 'Proper Motions', *IAU Colloq.* **7**, 143.

Penston, M. J.: 1973, *Monthly Notices Roy. Astron. Soc.* **164**, 121.

Pratt, N. M.: 1971, *Publ. Roy. Obs. Edinburgh* **8**, 109.

Purgathofer, A. Th.: 1969, *Lowell Obs. Bull.*, No. 147.

Walker, G. S.: 1971, *Publ. Roy. Obs. Edinburgh* **8**, 109.

PROPER MOTIONS OF OB STARS IN BOTH HEMISPHERES

HARUO YASUDA

Tokyo Astronomical Observatory, Japan

Abstract. From a comparison between the rotational velocities derived from radial velocities and space motions of OB stars, large systematic errors of FK4 proper motions in the southern hemisphere are evaluated; these may be expected, from the known accuracy of the FK4. The error of adopted distance scale is also examined. It is suggested that meridian observations of OB stars should be extended to the southern hemisphere to further researches, not only on stellar kinematics, but also on the fundamental system.

Gliese, Murray, and Tucker, 'New Problems in Astrometry', 227. All Rights Reserved.
Copyright © 1974 by the IAU.

ESTIMATION DE LA PRECISION DES MOUVEMENTS PROPRES DONNES DANS UN CATALOGUE D'ETOILES A L'AIDE DES OBSERVATIONS VISUELLES DE COUPLES STELLAIRES OPTIQUES

J. DOMMANGET

Observatoire Royal de Belgique, Belgium

Abstract. It is shown by means of several examples, that the visual observations of the relative rectilinear motion of components of wide optical double stars can be used to estimate the precision of catalogue proper motions.

Les nombreuses observations visuelles généralement disponibles pour tout couple stellaire dont la séparation atteint plusieurs dizaines de secondes d'arc (il s'agit alors généralement de couples optiques) permettent une bonne détermination du mouvement relatif de ses composantes. Le calcul, par la méthode des moindres carrés, de la trajectoire relative rectiligne, conduit à une précision généralement supérieure à celle des différences des mouvements propres donnés pour leurs composantes dans les catalogues stellaires de position, car les observations visuelles de ces couples sont non seulement assez nombreuses, mais couvrent souvent des intervalles de temps importants de l'ordre du siècle (figures 1a et 1b). Aussi, ces déterminations de trajectoires rectilignes devraient permettre de juger de la qualité des mouvements propres donnés dans les catalogues en comparant ceux-ci aux mouvements relatifs ainsi calculés.

Pour nous assurer des possibilités réelles de cette méthode, nous avons d'abord dressé la liste des couples pour lesquels une trajectoire précise avait été calculée (Dommanget et Nys, 1964) et pour lesquels aussi, les mouvements propres de chacune des composantes sont donnés dans trois catalogues: le 'Eigen-Bewegung Lexikon', le 'Catalogue de Yale' et le 'Catalogue du Smithsonian Astrophysical Observatory'.

Mais le matériel recueilli s'avérant par trop restreint (une dizaine de couples) – contrairement à nos espérances – nous avons calculé les trajectoires rectilignes relatives de quelques couples supplémentaires, intéressants de ce point de vue (Dommanget, 1969, 1974).

Le tableau I contient les dix-huit couples ainsi disponibles et donne, en plus des valeurs des mouvements propres relatifs de leurs composantes en ascension droite et en déclinaison, déduits des trajectoires, leurs écarts avec les valeurs correspondantes déduites des trois catalogues cités.

D'après les moyennes quadratiques de ces écarts, on constate que les valeurs tirées du EBL concordent mieux avec celles calculées par les trajectoires, que les valeurs provenant des deux autres catalogues et cela, tant en ascension droite, qu'en déclinaison, comme l'illustrent les graphiques de la figure 2.

Gliese, Murray, and Tucker, 'New Problems in Astrometry', 229–232. All Rights Reserved.

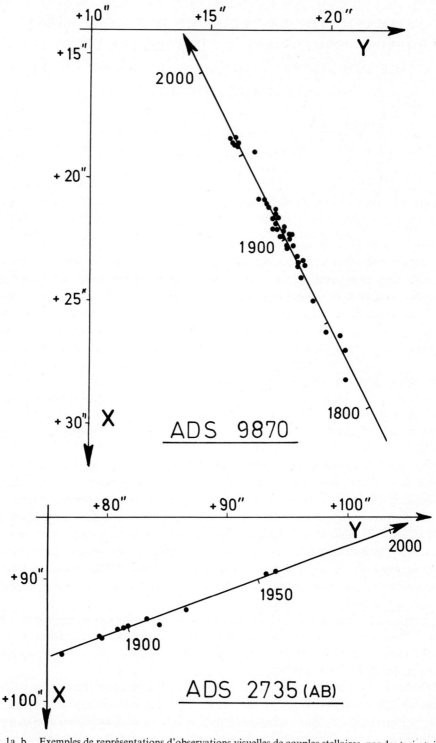

Fig. 1a, b.   Exemples de représentations d'observations visuelles de couples stellaires, par des trajectoires relatives rectilignes.

TABLEAU I

Etoile No. Cat. Index	Traj. rect.		EBL		Yale		SAO	
	α	δ	α	δ	α	δ	α	δ
00410 N 3024	$-0{,}''021$	$+0{,}''036$	$+0{,}''012$	$+0{,}''005$	$+0{,}''020$	$+0{,}''014$	$-0{,}''002$	$-0{,}''002$
00543 N 0015 (AB)	$+0{,}026$	$+0{,}102$	$-0{,}001$	$+0{,}016$	$+0{,}006$	$-0{,}005$	$+0{,}021$	$-0{,}006$
01437 S 0155 (AB)	$-0{,}006$	$-0{,}041$	$+0{,}001$	$-0{,}001$	$+0{,}024$	$-0{,}023$	$+0{,}024$	$+0{,}027$
02159 N 2310	$+0{,}048$	$0{,}000$	$+0{,}003$	$0{,}000$	$+0{,}012$	$+0{,}001$	$+0{,}007$	$-0{,}003$
03361 S 1256	$-0{,}094$	$-0{,}038$	$+0{,}002$	$+0{,}005$	$+0{,}040$	$+0{,}012$	$+0{,}014$	$+0{,}022$
03385 N 2735 (AB)	$+0{,}218$	$-0{,}081$	$-0{,}008$	$-0{,}027$	$-0{,}033$	$+0{,}013$	$-0{,}025$	$+0{,}028$
03594 N 2144 (AB)	$-0{,}165$	$+0{,}068$	$+0{,}003$	$+0{,}008$	$-0{,}008$	$+0{,}003$	$-0{,}008$	$+0{,}014$
04142 N 2707	$+0{,}007$	$+0{,}087$	$-0{,}001$	$+0{,}005$	$+0{,}025$	$-0{,}022$	$+0{,}001$	$-0{,}008$
05186 N 1716	$-0{,}230$	$-0{,}017$	$+0{,}011$	$+0{,}001$	$+0{,}022$	$+0{,}013$	$+0{,}003$	$+0{,}004$
06218 N 2051 (AB)	$+0{,}028$	$+0{,}034$	$+0{,}004$	$-0{,}001$	$+0{,}002$	$0{,}000$	$0{,}000$	$+0{,}007$
07532 N 0229 (AB)	$+0{,}173$	$-0{,}110$	$-0{,}010$	$+0{,}004$	$-0{,}028$	$+0{,}003$	$-0{,}027$	$-0{,}001$
09029 N 2702 (AB)	$+0{,}115$	$+0{,}324$	$+0{,}012$	$-0{,}002$	$+0{,}015$	$+0{,}009$	$+0{,}034$	$+0{,}004$
09094 N 2348 (AC)	$+0{,}138$	$+0{,}088$	$+0{,}006$	$-0{,}029$	$+0{,}001$	$+0{,}001$	$-0{,}012$	$-0{,}021$
09488 N 0525	$-0{,}020$	$-0{,}030$	$+0{,}003$	$+0{,}002$	$+0{,}004$	$-0{,}016$	$-0{,}011$	$-0{,}010$
13097 S 1050 (AB)	$+0{,}221$	$+0{,}311$	$-0{,}012$	$+0{,}009$	$-0{,}037$	$+0{,}020$	$-0{,}014$	$+0{,}003$
14516 S 2058 (AB)	$-0{,}058$	$+0{,}082$	$-0{,}032$	$+0{,}008$	$+0{,}035$	$-0{,}020$	$-0{,}011$	$-0{,}010$
19202 N 1143 (AB)	$-0{,}679$	$-0{,}707$	$-0{,}012$	$-0{,}022$	$-0{,}002$	$-0{,}024$	$-0{,}014$	$-0{,}032$
20264 N 1055 $(A-BC)$	$-0{,}019$	$-0{,}005$	$-0{,}003$	$-0{,}005$	$+0{,}009$	$-0{,}012$	$+0{,}005$	$-0{,}003$
Moyennes quadratiques			$\pm 0{,}''011$		$\pm 0{,}''019$		$\pm 0{,}''016$	

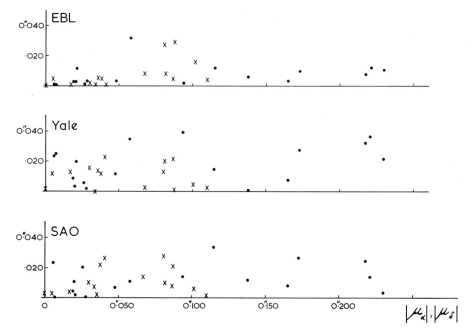

Fig. 2. Représentation graphique, pour les 18 couples stellaires considérés ici, des écarts entre les mouvements propres relatifs déduits des catalogues EBL, de Yale et du Smithsonian Astrophysical Observatory et ceux déduits des trajectoires rectilignes calculées à partir des observations visuelles. Les signes de ces écarts ainsi que ceux des valeurs de μ_α et de μ_δ ont été ignorés. Les points correspondent aux écarts en α, les croix aux écarts en δ.

Remarquons que tous les couples retenus présentent une distance de séparation comprise entre 15″ et 100″ environ, ce qui permet d'éviter d'une part les effets dûs à la proximité des images photographiques stellaires et, d'autre part, les trop grandes imprécisions entachant les mesures visuelles des couples très écartés. Par ailleurs, les observations couvrent, dans chaque cas, au moins des intervalles de temps de près d'un siècle.

Il n'empêche qu'avant de pouvoir tirer une conclusion certaine d'une telle comparaison, il y aurait lieu d'entreprendre le calcul des trajectoires relatives d'un plus grand nombre de couples pour lesquels les mouvements propres des composantes sont donnés dans les catalogues à comparer.

Aussi, des trajectoires rectilignes sont régulièrement calculées à l'Observatoire Royal de Belgique, dans ce but.

Pour terminer, nous insisterons sur le fait que la prétention de cette note se limite à montrer les possibilités de la méthode envisagée pour évaluer la qualité relative des mouvements propres donnés dans les catalogues d'étoiles.

Bibliographie

Dommanget, J.: 1969, *Bull. Astron. Obs. Roy. Belgique* **6**, 343.
Dommanget, J.: 1974, *Bull. Astron. Obs. Roy. Belgique* **8**, sous presse.
Dommanget, J. et Nys, O.: 1964, *Ann. Obs. Roy. Belgique, 3ème Sér.* **9** (6), 309.

ASTRONOMICAL REFRACTION PROBLEMS

THE PRESENT STATE AND FUTURE OF THE
ASTRONOMICAL REFRACTION INVESTIGATIONS*

G. TELEKI

Astronomical Observatory, Belgrade, Yugoslavia

Abstract. (1) Astronomical refraction is one of the oldest astronomical problems, which has been investigated many times, but the existing knowledge of refraction is not sufficient to meet modern requirements for accurate astrometric data. The present investigations cast light on several facts which might help the advancement in the calculation of astronomical refraction.

(2) There are several questions concerning the calculation of the astronomical refraction, from the laws of refraction to meteorological factors. These difficulties have prevented the development of a satisfactory refraction theory for a long time, leading to the conservation of old methods, but the accuracy of the refraction determinations has not increased. We are using old refraction theories and tables based on them for the most part, while the anomalous refraction is not considered. Seeing that the calculated refraction values are not acceptable, we are determining the formal refraction corrections directly from astrometric measurements, but these corrections are not real from the physical point of view.

The present uncertainty in the calculated refraction is of the order of $0\rlap{.}''01$, increasing with zenith distance. For observations near the horizon the existing refraction theories cannot give acceptably accurate values.

(3) The greatest progress in the knowledge of refraction influences would be achieved if instead of mean meteorological data we could use direct measurement. There are some propositions of this kind.

The new refraction tables are in preparation. We are very careful in taking into consideration the local characteristics and the correct calculation of the chromatic refraction.

Frankly speaking, we must say that it is practically impossible to calculate a real influence of refraction and for this reason certain precautions are advisable. A careful positioning of the instrument may simplify the density field and thus eliminate the greatest part of the anomalous refraction. But the best proposition is to observe at small zenith distances only, not larger than $45°$ if possible.

(4) Our conclusion is that, in current astrometric practice, we must continue refraction investigations and in addition to the refraction at optical wavelengths, it is absolutely necessary to investigate the radio and laser-refraction too.

---

* The full text of this paper will appear in *Publications de l'Observatoire Astronomique de Béograd*, No. 18.

DISCUSSION

Van Herk: The geodesists suffer from difficulties with refraction as well as the astronomers do. They have formed a study commission with, as co-ordinator, Dr E. Tengström, Geodetic Institute, Uppsala University, Hällby, Uppsala 75590, Sweden.

Maybe the two study commissions could aid each other.

Teleki: I will propose the co-operation.

Strand: Some 20 years ago refraction at large distances was measured at the Dearborn Observatory from separation of craters on the moon, as the moon was rising over Lake Michigan. Comparison with the theoretical refraction and using available atmospheric data, showed clearly that the actual refraction could differ as much as 5% from the computed value.

Teleki: It can be seen that there are many problems connected with the refraction.

Kovalevsky: It is to be remarked that theoretical and observational material analysed by Miss Hopfield showed that the dry component of refraction in distance (for laser measurements) can be modelled from observations of temperature and pressure at the site of the observer to an accuracy of about 1 cm.

ENVIRONMENTAL SYSTEMATICS AND ASTRONOMICAL REFRACTION, I*

J. A. HUGHES

U.S. Naval Observatory, Washington, D.C., U.S.A.

Abstract. The first results of a planned continuing investigation of astronomical refraction are reported in this paper. Pure and anomalous refraction are investigated using meteorological data and environmental models. Numerical methods based upon Chebyshev polynomials are developed. Applications are made to Winslow, Arizona (which is somewhat similar to El Leoncito, Argentina), and to results of PZT observations made in Washington, D.C.

DISCUSSION

Van Herk: The results for the constant of aberration from the PZT in Washington have always differed systematically from the one derived in Richmond. Could this yearly refraction term you mention be the cause of this discrepancy?

Teleki: Frankly speaking I don't know.

Klock: I would like to mention that Dr Hughes was encouraged to undertake this investigation by Mr F. P. Scott, who, after examining the first results from the El Leoncito Station, realised that the Pulkovo refraction table did not appear to cover the El Leoncito situation appropriately.

---

* The full text of this paper will appear in *Publications de l'Observatoire Astronomique de Béograd*, No. 18.

LES EFFETS DE LA REFRACTION ATMOSPHERIQUE SUR LES COORDONNEES TANGENTIELLES EN ASTROMETRIE

J. DOMMANGET

Observatoire Royal de Belgique, Belgium

Résumé. On montre que les coordonnées tangentielles considérées en astrométrie et affectées de la réfraction, peuvent toujours s'écrire, même pour des champs pouvant s'étendre théoriquement jusqu'à 90°, sous la forme:

$$X_r = X + aX_a + R_x$$
$$Y_r = Y + aY_a + R_y,$$

où a est le coefficient du premier terme de l'expression de la réfraction:

$$r = a \operatorname{tg} \xi + b \operatorname{tg}^3 \xi ;$$

X_a et Y_a, des expressions calculables aisément pour chaque point du cliché et R_x et R_y, les restes des développements des coordonnées réfractées X_r et Y_r. Ces restes sont fonctions de a et de b théoriquement, mais leur calcul pour chaque point du cliché peut se faire en toute rigueur même avec des valeurs approchées de ces coefficients. Introduits sous cette forme dans les formules de réduction des clichés, les effets de la réfraction atmosphérique n'apparaissent que par le seul paramètre inconnu a, et de manière linéaire.

ASTROMETRIC TECHNIQUES

MODERN DEVELOPMENTS OF THE MERIDIAN CIRCLE

(Invited Paper)

ERIK HØG

Hamburger Sternwarte, F.R. Germany

Abstract. The successful improvements of meridian instrumentation during the past twenty years have mainly come through electronic devices and computers being applied for facilitating the data handling and for increasing the accuracy. Photographic recording of the circle and of the star has played an important role here, but has for both tasks gradually lost its attractiveness as direct photometric-electronic methods have become available. At the same time new types of the telescope system have been introduced without convincing results as yet, but this field still holds promises for the near future.

The problems of the refraction, the meridian building, the foundation of the instrument and the site selection have been treated in recent years and these efforts will be doubly repaid when the telescope and micrometers become more nearly perfect. Altogether, it need not be long before a few partly automatic instruments produce observations with a mean error of 0″.15 and a systematic error of 0″.03 at a rate of 300 observations per observing night. In addition the limiting magnitude can be $m_v = 11$ or 12, thus 2 mag. fainter than for visual observations. These goals are conservative – most of them have already sometimes been surpassed – and they should be compared to a present day good visual meridian circle giving a mean error 0″.30, a systematic error 0″.10 and 120 observations per night.

The relative roles of meridian and photographic astrometry must be defined anew in the light of the great improvements of both methods.

1. Introduction

The opinion sometimes expressed that the accuracy of meridian observations improves very slowly, can no longer be maintained. This opinion was well supported by evidence since the mean error of a single observation was about 0″.45 around 1890 and is 0″.30 for visual observations today, and since the systematic errors have hardly improved at all since then (van Herk and van Woerkom, 1961; Gliese, 1965; Gauss, 1971).

About ten years ago, however, it was obvious that the state of art of available technology would permit photographic and photoelectric micrometers to be built which would reach nearly to the limit of accuracy set by the disturbances from the atmosphere. Corresponding successful developments of three different methods were started at the Brorfelde, Bergedorf, and Bordeaux observatories which have now given about 160000 observations with an average mean error of 0″.22. It is equally obvious that further improvements of the accidental errors and even more of the systematic errors can be obtained. These improvements will – of course – not come by themselves but only if decisions are taken to support such work. It carries good promise for the future that new meridian circles employing modern techniques are being put into operation at Caracas, Sao Paolo and Washington, and one is planned for Tokyo. Regrettably, at the same time at some other places the support has been drastically reduced.

The last twenty years have brought such improvements of the circle, its reading

Gliese, Murray, and Tucker, 'New Problems in Astrometry', 243–255. All Rights Reserved.

and the determination of the division errors that these subjects impose no limitation on the observing accuracy or working efficiency today.

Prior to this recent period went a time when meridian astronomers started utilizing electronic computers as soon as they became available to relieve the meridian observer from the burden of routine tasks. Watts (1960) and Adams (1963) described the pioneering work at the U.S. Naval Observatory. The introduction of quartz clocks and a digital chronograph at R.G.O. Herstmonceux has been surveyed by Tucker (1963).

Thus, electronic equipment and computers have been at the center of the field, indicating that the progress will continue to be fast. On-line computers are just now starting to be employed for data-acquisition and for the control of micrometers and telescope setting. They will radically decrease the amount of routine work still involved with the operation of the recent photographic and photo-electric micrometers.

In Table I performance data of some modern meridian circles are given. The joint

TABLE I

Performance of modern meridian circles

Meridian circle	Type of micrometer	Observations				Mean errors		Arc	
		Start	Years	Nights	Obser-vations	$\Delta\alpha\cos\delta$	$\Delta\delta$	in α	in δ
11 m.c.'s AGK3R	visual	1956	6	–	300 000	0^{s}016	0″35	–	–
In Bergedorf	visual	1956	6	360	41 611	0.016	0.42	2^h	60°
In Perth	MSM[a]	1967	5	580	110 000	0.012	0.27	7	140
In Brorfelde	photogr.	1964	8	300	50 000	0.015	0.22	2	25
In Bordeaux	tracker	1971	–	34	1 500	0.007	0.20	4	80

[a] Multislit micrometer.

international enterprise of AGK3R, of which the contribution of the 19 cm visual Repsold instrument in Bergedorf is a part, is given as basis for a comparison. The Table gives (2) the type of micrometer, (3) the start of the observations, (4) the number of years observed, (5) the number of nights, (6) the number of observations, (7) and (8) the mean errors of one observation, (9) and (10) the arc on the sky which was tied to the FK4-stars as one unit.

The photoelectric instrument in Perth has acquired observations with a statistical weight equivalent to 200 000 visual observations although for technical reasons one fourth of the available clear nights were not used. During the preceding 60 years in Bergedorf 130 000 observations were obtained.

The systematic errors are determined by many factors other than the micrometers and the circle: telescope, pivots, stability of foundation, and refraction inside and outside the dome. Improvements here have not yet been so obvious.

The publications on all aspects of meridian techniques found in: *Trudy Astrometricheskoi Konferenzii U.S.S.R.*, *Bulletin of Astron. Obs. Pulkovo* and *Soviet*

Astronomy (=*Astronomicheskii Zhurnal*) constitute an inexhaustible source of inspiration for the meridian astronomer, even in the cases where he can only read the summary in *Astronomischer Jahresbericht*. Since they are too numerous to mention completely in this limited space, they deserve to be quoted here as an entity prior to the following detailed discussion.

2. Micrometer

Visual micrometers of the travelling wire type are still widely used by meridian observers but cannot in the long run compete with some of the new photographic and photoelectric micrometers with respect to accuracy, limiting magnitude and ease of operation and evaluation.

Performance data for new operating or proposed micrometers are collected in Table II as they are actually obtained or expected under average seeing conditions

TABLE II

Performance of meridian circle micrometers under comparable conditions

Micrometer	m.e. (in R.A.)	Limit m_v	Cathode	Autocol- limation	Daytime objects
Visual	0″.24	10	–	yes	yes
Photogr., Brorfelde	0.22	11	–	no	no
MSM, Perth	0.18	10.5	'S'	yes	–
MSM, proposed	0.15	11.5	'S'	yes	yes
Tracker, Bordeaux	0.10	9	'S'	no	no
Tracker, Klock	–	10.5	S20	yes	(yes)
Optimum photoelectric	0.13	15.2	'S'	yes	yes

at a 20 cm refractor and with an observing time $T=40$ s. The mean error for R.A. is given since the accuracy of a micrometer in Decl. is mostly deteriorated by the errors of the circle reading. For the optimum micrometer defined in Section 2.3, the mean error due to image motion alone is derived from the formula (Høg, 1968)

$$\sigma_T = 0″.33\,(T+0.65)^{-0.25} \tag{1a}$$

valid for 0.2 s $\leqslant T \leqslant 14\,000$ s and at the zenith. This is equivalent to the power spectrum

$$P(f) = 0.08\,f^{-0.5}(1+2f)^{-1}\,\square''/\text{cps} \tag{1b}$$

for the frequencies 0.00001 cps $\leqslant f \leqslant 10$ cps. It must be due to a good micrometer and to an unusually small image motion in Bordeaux that this accuracy has been surpassed by Requième (1973).

The limiting magnitudes discussed below are reduced to a visual wavelength region, $\lambda > 5100$ Å, which is most important in order to minimize the difference in Decl. for different spectral types due to the atmospheric dispersion. The use of a S20 photocathode instead of a 'S' cathode would bring a gain of one magnitude provided the higher dark currents from cathode and sky can be coped with. It is worth noting that the photographic and the multislit micrometers give the faintest stars, contrary to the theoretical expectation that a tracker should utilize the light more efficiently (see Section 2.2 and Siedentopf, 1963).

Measurement of autocollimation onto a mercury mirror is important for the physical determination of the instrumental constants as needed when other than differential observations relative to an existing fundamental catalogue are wanted. Such autocollimation measurements are always possible with visual micrometers but only with the tracker of Klock (1970) and with the MSM (multislit micrometer), see Table II.

Daytime observations of the Sun, the inner planets and stars are required for absolute observations of the zero-points of the celestial coordinate system. Again, besides visual micrometers, only Klock's tracker (at a later stage of development) and the MSM will be capable of doing this.

2.1. PHOTOGRAPHIC MICROMETERS

Photographic micrometers have long been used at photographic zenith tubes and the same principle of moving the plate along with the star has been transferred to the meridian circle in Brorfelde by Laustsen (1967). This micrometer is convenient to operate for a single observer, it gives a mean error of $0''.22$ in both coordinates. No measurement of autocollimation or daytime objects is possible. The measurement of the photographic plates constitutes a bottle neck, although a digitized manual measuring machine is used. A number of plates have been measured with the automatic measuring machine GALAXY but without an encouraging success although GALAXY performed perfectly well. The task of identifying the measured images with the images on the plate caused problems (Fogh Olsen, private communication).

Although vertical circles are outside the scope of this paper we mention the photographic micrometer of the Pulkovo vertical circle described by Zverev et al. (1966), and at the present symposium by Bagildinskij. By observation of star trails a mean error of $0''.25$ for one absolute observation of Decl. is obtained.

2.2. PHOTOELECTRIC MICROMETERS

Photoelectric micrometers have a long history in meridian instrumentation, starting with Strömgren's proposal in 1933 of a slit micrometer. But the early ideas had to await the advent of the photomultiplier, digital electronic equipment and of the digital computer before they could bear efficiently in practice. Pavlov (1956) has used a slit micrometer with a reflecting grid for regular transit observations and has obtained over 80000 observations of R. A. (Pavlov et al., 1971).

A photoelectric multislit micrometer for simultaneous observation of R.A. and Decl. described by Høg (1970, 1972a) has been used in Perth since 1967. The faintest stars in the SRS catalogue were of $m_v = 10.5$ and no fainter stars could have been set by the observer in the average night although they could have been measured.

Measurements of autocollimation onto the nadir mirror were obtained every 3 or 4 hours. Daytime objects were not observed, mainly due to lack of time.

An improved multislit micrometer has been proposed (Høg, 1972b). The new grid is not fixed as in the previous micrometer but must be mounted on a stage with a very accurate motion in the direction of R.A. This motion will give three advantages:

(1) Stars very near to the pole (polarissimae) can be observed in the same time as other stars when a scanning motion of the grid is used.

(2) An automatic star acquisition (after a coarse setting in declination) is provided so that one of the two observers required with the old micrometer is no longer needed.

(3) The autocollimation measurement is simplified. Thus, the draw-backs of the old micrometer will be removed. Observation of all daytime objects including Venus will be possible.

The use of a photoelectric star sensor or star tracker is the most direct way to replace the human eye in the impersonal Repsold micrometer. An account of the many different types of star sensors has been given by Kühne (1971) and only a few names shall be mentioned here: Rotating knife-edge, quadrant sensors with and without chopper, tridrant sensor, frequency method, counting method.

The tracker for the Automatic Transit Circle at Washington (Klock, 1970) has promising universal capabilities as stated above.

Requième (1973) presents results from a rotating knife-edge micrometer at the 19 cm meridan circle at Bordeaux. A limiting magnitude of at least 10^m is expected when the sky background is reduced but only using a large wavelength region down to about $\lambda 4100$.

The tridrant star sensor with chopper developed by Kühne (1971) has a limiting magnitude $8^m\!.0$ with a 15 cm refractor, a S20 cathode and without color filter.

A limitation of the color regions for the sensor of Requième and Kühne to $\lambda > 5100$ Å will bring a loss of at least one magnitude.

The somewhat disappointing sensitivity of star trackers may in principle be improved by increasing the time constant of the detecting system (about 1 s in Requième's tracker), by using sensor systems with more than one photo-detector (quadrant sensor) and by improved photometric evaluation with an on-line digital computer. A (remote) design goal of an optimum micrometer is presented below.

2.3. LIMITING MAGNITUDE

For comparison with real micrometers we introduce the concept of the optimum photoelectric micrometer which must be able to integrate in a perfectly linear way the two-dimensional intensity distribution of the star image during the observing time. A subsequent evaluation shall give the mean position of the image. A photographic plate moved along with the star is an approximation to this optimum concept but has the disadvantages of a non-linear integration and a relatively low quantum efficiency. For the optimum micrometer we suppose the same quantum efficiency as available 'S' photocathodes, (5% at $\lambda 5500$), although S20 cathodes would bring a gain of at least one magnitude for the visual region.

The limiting magnitudes which can be obtained with different micrometers under comparable conditions are given in Table II. For the trackers a loss of one magnitude has been supposed when only the visual wavelength region is used as for all the other micrometers. The values for the proposed MSM and for the optimum micrometer have been obtained using the realistic image profiles $i'_{star}(x)$ shown in Figure 1a measured with a very narrow slit (Høg, 1971b). The counting rate for $m_v = 11.5$ has been supposed to be $i_{star} = 100$ c s^{-1} as is obtained with a 'S' cathode. For a slit of width $s_w = 4$ (see Figure 1b) the signal $i_s(x, s_w)$ is shown in Figure 1c. Using the median method to derive the transit time for both the MSM and the optimum micrometer similar deductions as used earlier (Høg, 1970) give the shot noise at a single slit to

$$\sigma_{shot} = v(2bi_d + s_t i_{star})^{0.5}/(2i_{slit}) \tag{2}$$

and the shot noise for the optimum micrometer to

$$\sigma_{shot} = (Ti_d + Ti_{star})^{0.5}/(2i'_{star}T), \tag{3}$$

where $v['' / s]$ is the speed of the star, $b = 6''.6/(v \cos 45°)$, $s_t = s_w/(v \cos 45°)$ and T the total observing time. The dark current i_d from sky and photocathode is nearly negligible for a dark sky of 130 $\square''$ area and a 1 mm cathode.

It is evident from Table II that all photoelectric micrometers are very far from an optimum utilization of the photons even though a low quantum efficiency has been supposed for the optimum micrometer.

3. Declination Circle

The classical declination circle consists of equidistant narrow lines every few minutes of arc which are measured by four or six equidistant microscopes. The microscopes must be moved to other positions on the periphery for determination of division errors. Other types of circles exist but an analysis shows that for different reasons none of them are complete alternatives for a meridian circle. E.g. the Inductosyn used by Klock et al. (1970) has the handicap that its division errors can only be determined by means of another classical circle on the same axis.

Four problems of the classical declination circle may be distinguished:
(1) The quality of the circle, be it of metal or of glass.
(2) The illumination of the circle to provide lines of high contrast.
(3) The recording and measuring, be it photographic or photoelectric.
(4) The division corrections.

3.1. Circle Illumination

Some modern meridian instruments have a circle consisting of a massive glass ring which carry photochemically produced division lines. Lines on glass must be illuminated *from behind* in order to create a high contrast which is essential for repeatable readings. Einicke et al. (1971, p. 13) obtained in this way and with photoelectric

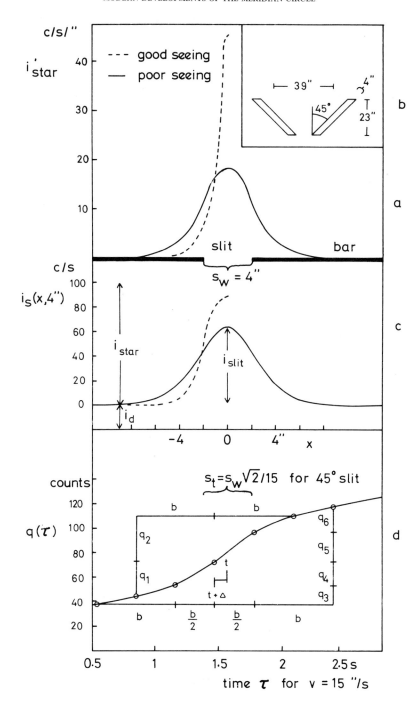

Fig. 1. (a) Realistic smoothed profiles of focal images ($m_v = 11.5$), (b) slits of proposed multislit micrometer, (c) realistic smoothed photometric curves from this micrometer, (d) integrated photometric curve (for poor seeing), cf. Høg (1970, p. 94).

scanning micrometers the best accuracy ever seen: 0".03 as the mean error for reading at six microscopes (including the errors of the division corrections).

Older meridian instruments have steel circles where the lines are engraved in a layer of gold and filled with a dark material. Such lines do not have a high contrast, and the mean error of a reading on four microscopes even with scanning micrometers was only 0".15 (Høg, 1972a). Since it is difficult to replace the metal circle on an old instrument with a glass circle, it is important to note the existence of two new methods: (1) to divide metal circles with very sharp lines of high contrast as offered by Société Genevoise d'Instruments de Physique, and (2) to mount a very thin glass circle directly on the old circle as offered by Teledyne Gurley, New York.

The new metal circle divisions are engraved on a specular reflecting layer of nickel on a plane steel base, but they are not filled with dark material. This specular reflecting surface gives a higher contrast to the lines than the old gold circles which had a partly diffuse and a partly specular reflection. The illumination system must be carefully designed to avoid vignetting, which easily arises when illuminating a specular reflecting surface, and which would cause serious systematic errors.

3.2. Circle reading

Both for reasons of economy and accuracy the direct photoelectric scanning of the circle with a slit is preferable to the widely applied photographing and subsequent measurement in the laboratory. Scanning is now used at the observatories of Brorfelde, Perth and Washington either with output of all photometric values on punched tape or, more recently, onto a small on-line computer, thus avoiding the kilometers of tape formerly needed.

3.3. Division corrections

The determination of division errors of every single diameter is required nowadays. Formerly, with visual reading and laborious methods of reduction this could take about 10 man-years. With the general symmetric method (Lévy, 1955; Høg, 1960b) and scanning micrometers it was done in one man-year for two circles with 3' divisions (Fischer-Treuenfeld, 1968) and in a man-month (Einicke *et al.*, 1971) for one circle with 5' divisions. This year it has been done in Brorfelde with an on-line computer in one man-week! (Fogh Olsen, private communication.) Many irregular errors in observations of declinations and also systematic errors $>0".1$ over large arcs (Einicke *et al.*, 1971, p. 23) may be explained by incomplete knowledge of the division errors which were often determined only for the half-degree lines.

4. Telescope System

If we look back a decade to van Herk's and van Woerkom's account (1961) of the problems in meridian astronomy it is noteworthy that good solutions have been developed for the problems of the micrometer and of the circle and its reading. But we should still look carefully at their discussion of the telescope and the local meteorological effects.

The photoelectric micrometer seems to stabilize the telescope itself as the expected consequence of the much smaller thermal influence of the observer on the telescope. This is concluded from the small clamp differences in R.A. of the Perth observations (Høg and Nikoloff, 1973) which show that internal agreement within 0".02 can be obtained provided the pivots are good enough.

I do therefore think that classical and even old meridian circles can perform excellent work if the micrometers, the circle, the pivots and the dome are brought up to a modern standard, including an exact determination of pivot and circle errors. The development of a new telescope system is not the most urgent task – but surely the most difficult.

One of the strongest reasons to develop a new telescope system is the desire to eliminate the flexure of the rotable telescope tube, a problem which has been discussed by many authors (Atkinson, 1955). Also the small uncontrolled shifts of the lens elements need a new solution (Orlow, 1953; Naumov, 1966). The predictions in this paper of the future accuracy of meridian circles are based on the use of conventional instruments and do not include the possibly very great improvements from new telescope systems.

Three new types of meridian circles are depicted in Figure 2 together with the conventional and shall be discussed here.

Development of the horizontal meridian circle (c) which carries the names of R. d'E. Atkinson and of L. A. Sukharev has been pursued at the observatories of Greenwich, Ottawa, Oporto and Pulkovo, but has now been abandoned for different reasons at the three first places. The most recent results from testing the instrument in Pulkovo with which only R.A. can be measured are given by Pinigin (1972 and private communication) who states that the mean error of one photoelectric auto-collimation reading is about 0".01 and of one determination of Bessel's n is $\pm 0^s.010$. 3000 visual observations of stars have been obtained with a mean error $0^s.011$ sec δ and they are essentially free from systematic errors. The instrument has only very small variations with time and temperature.

A cassegrain type automatic mirror transit circle, ATC, is being tested at the U.S. Naval Observatory by Klock (1970, 1973).

A horizontal glass meridian circle, GMC, has been proposed by Høg (1971a).

A direct comparison of these three instruments is impossible since they are in completely different stages of development but some basic problems seem worthwhile to point out.

Six specific problems exist for the Atkinson type instrument most of which have already been pointed out by Atkinson: (1) refraction and seeing disturbances in the long horizontal light paths, (2) a tilt-free connection within 0".01 between the mirror and the axis and the circle, (3) two telescopes are needed to cover all declinations, (4) these horizontal telescopes tend to obstruct the line of sight to any possible azimuth marks, (5) different parts of the mirror are used at different declinations and (6) the circle must be read with twice as high an accuracy.

The supporting system for the mirror tested by Atkinson (1961) was shown to

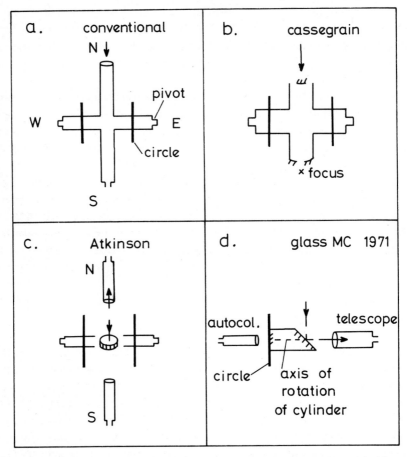

Fig. 2.   Four types of meridian circles all pointed at north and viewed from the zenith: (a) rotable refracting telescope, (b) rotable cassegrain telescope, (c) rotable mirror and two fixed horizontal telescopes,
(d) rotable zero-expansion glass cylinder.

solve the second problem in R.A. while no direct check of the tilt in Decl. at different
zenith distances is possible. In the Pulkovo mirror transit the mirror and the axis is
one piece of steel, thus solving radically problem No. 2.

A specific problem of the new ATC in Washington seems to be that the optical-
mechanical system is very complicated. This is certainly a danger for the economy
and possibly a danger for the accuracy and the reliability. Its automatic operation
under computer control is an important step forward and will help to acquire a
wealth of astrometric information.

Four specific problems of a GMC in the new version, Figure 3, have been pointed
out (Høg, 1972c) and concern the support and bearing for the glass cylinder, the
horizontal refraction and the quick measurement of autocollimation. The principal
advantages of a GMC are: (1) very small flexure, (2) relative simplicity, (3) it requires
a much smaller building than a conventional meridian circle (Høg, 1973).

5. Foundation, Building, Site

The performance of a meridian circle depends on its surroundings, the more so as the instrument becomes more perfect. Since this concern will find expression in several other lectures of this symposium we can be brief here.

A discussion has recently been given elsewhere (Høg, 1973) about refraction anomalies and of the foundation and its possible improvements.

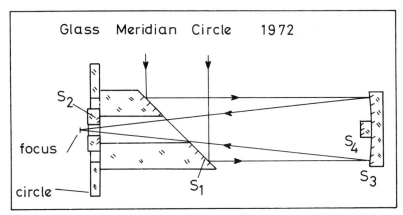

Fig. 3. The Glass Meridian Circle in the new version where the fixed telescope and the autocollimator have been combined in the mirror S_3. The mirror S_4 is cemented to S_3 and since its center of curvature coincides with the focus of S_3, it can serve to control the tilt of S_3.

The building protects the instrument when it is not used, but should, ideally, be completely removed during observing in order not to create local heating of the instrument and atmosphere. This ideal can be most nearly reached if the instrument is so constructed that no protection against the wind is required and only a small building is needed. The ideal has been approached with success for the transit instrument by Pavlov (1963, 1972), while the bulky structure of the conventional type of meridian circle and the Atkinson type prevents much progress in this respect. Promising is the pavilion for the Glass Meridian Circle as shown by Høg (1972c, 1973).

The selection of a site especially for an astrometric observatory has been undertaken for the first time in France as described by Laclare (1969) and Kovalevsky (1972). The very slow parts of atmospheric image motion extending over minutes and hours of time are of primary relevance to astrometric observations whereas they hardly influence other astronomical observations. Therefore, observations with the Danjon astrolabe were used to measure the stability of the images. The best stability was found at stations where air masses of marine origin dominate.

6. Conclusion

With the implementation of available meridian techniques absolute positions of faint stars of $m_v = 11$ and 12 can soon be obtained with an accuracy of 0.″05 from observations with a single *conventional* meridian circle. Such observations of reference stars for long-focus astrometric measurement of optical positions of radio sources would bring a corresponding improvement of the accuracy of $\pm 0.″15$ obtained by Murray *et al.* (1971). The accuracy of absolute radio positions is about $\pm 0.″5$ (Fricke, 1972) and is gradually improving towards the limit now reached in optical meridian astrometry where the atmosphere becomes the most important disturbance, which will therefore contribute similarly to the errors of optical and of radio positions.

This common limitation and the peculiar characteristics of the power spectrum of image motion (Høg, 1968) are mostly forgotten when a positional accuracy of 0.″001 is predicted for radio astrometry. It is a challenge to meridian astrometry to maintain its lead concerning the accuracy of absolute positions, but in any case the two techniques must complement each other as far as they are concerned with different celestial objects.

The future will see fewer meridian circles acquire an increasing number of more accurate positions at a lower cost thanks to the automation. At the same time photographic observations will be available with improved astrographs (de Vegt, 1973) of the 2 m focal length class measured with automatic machines and reduced with the overlap-technique. It is important to discuss anew the most efficient combinations of the improved meridian and astrograph techniques for the many different astronomical purposes.

References

Adams, A. N.: 1963, *Symp. über Automation...*, Heidelberg 1963, p. 88.
Atkinson, R. d'E.: 1955, *Monthly Notices Roy. Astron. Soc.* **115**, 427.
Atkinson, R. d'E.: 1961, *Roy. Obs. Bull.*, No. 34.
De Vegt, Chr.: 1973, this volume, p. 209.
Einicke, O. H., Laustsen, S., and Schnedler Nielsen, H.: 1971, *Astron. Astrophys.* **10**, 8.
Fischer-Treuenfeld, W. F. von: 1968, Dissertation, Hamburg.
Fricke, W.: 1972, *Ann. Rev. Astron. Astrophys.* **10**, 101.
Gauss, F. S.: 1971, *Astron. J.* **76**, 492.
Gliese, W.: 1965, *Astron. J.* **70**, 162.
Høg, E.: 1968, *Z. Astrophys.* **69**, 313.
Høg, E.: 1970, *Astron. Astrophys.* **4**, 89.
Høg, E.: 1971a, *Mitt. Astron. G.* **30**, 148.
Høg, E.: 1971b, *Astrophys. Space Sci.* **11**, 22.
Høg, E.: 1972a, *Astron. Astrophys.* **19**, 27.
Høg, E.: 1972b, available on request: 'Design Study of a Multislit Micrometer and a Semi-automatic Meridian Circle'.
Høg, E.: 1972c, *Mitt. Astron. G.* **32**, 120.
Høg, E.: 1973, *Mitt. Astron. G.*, in press.
Høg, E. and Nikoloff, I.: 1973, this volume, p. 79.
Klock, B. L.: 1970, *Trans. IAU* **XIVA**, Report Commission 8 (mimeographed).
Klock, B. L., Geller, R. Z., and Dachs, M. A.: 1970, *U.S. Naval Obs.*, Reprint No. 107.
Kovalevsky, J.: 1972, *Mitt. Astron. G.* **31**, 49.
Kühne, C.: 1971, *Mitt. Astron. G.* **30**, 109.
Laclare, F.: 1969, Dissertation, Paris.
Laustsen, S.: 1967, *Publ. Mind. Medd. Kbh. Obs.*, Nr. 190.
Lévy, J.: 1955, *Bull. Astron.* **20**, 35.
Murray, C. A., Tucker, R. H., and Clements, E. D.: 1971, *Roy. Obs. Bull.*, No. 162.

Naumov, V. A.: 1966, *Bull. Astron. Obs. Pulkovo* **24**, 22.
Orlow, B. A.: 1953, *Bull. Astron. Obs. Pulkovo* **19**, 56.
Pavlov, N. N.: 1956, *Trudy 13th Astrometr. Konf. U.S.S.R.*, p. 64.
Pavlov, N. N., Afanasjeva, P. M., and Staritsyn, G. V.: 1971, *Bull. Astron. Obs. Pulkovo, 2nd Ser.* **1** (78), 27.
Pavlov, N. N.: 1963, *Trudy 15th Astrom. Konf. U.S.S.R.*, p. 265.
Pavlov, N. N.: 1972, *Trudy 18th Astrom. Konf. U.S.S.R.*, p. 158.
Pinigin, G. I.: 1972, *Trudy 18th Astrom. Konf. U.S.S.R.*, p. 165.
Podobed, V. V.: 1962, *Fundamental Astrometry* (English translation 1965).
Requième, Y.: 1973, *Astron. Astrophys.* **23**, 453.
Siedentopf, H.: 1963, *Symposium über Automation...*, Heidelberg 1963, p. 57.
Strömgren, B.: 1933, *VJS der AG* **68**, 365.
Tucker, R. H.: 1963, *Symposium über Automation...*, Heidelberg 1963, p. 112.
Van Herk G. and van Woerkom, A. J. J.: 1961, *Astron. J.* **66**, 87.
Watts, C. B.: 1960, *Stars and Stellar Systems* **1**, 80.
Zverev, M. S.: 1954, *Fundamental Astrometry* (English translation 1963).
Zverev, M. S., Naumova, A. A., Naumova, V. A., and Torres, K.: 1966, *Bull. Astron. Obs. Pulkovo* **24**, 12.

DISCUSSION

Klock: I would just like to mention my concurrence with Dr Høg on many of the ideas in his paper and particularly note his foresight in the close race being given to us by the radio astrometrist, as evidenced by the papers presented here yesterday.

INSTRUMENTAL PARAMETERS OF THE U.S. NAVAL OBSERVATORY'S AUTOMATIC TRANSIT CIRCLE (ATC)

B. L. KLOCK and F. S. GAUSS

U.S. Naval Observatory, Washington D.C., U.S.A.

Abstract. The instrumental constants of the Naval Observatory's Automatic Transit Circle are discussed in conjunction with the transit circle's on-line data acquisition and control system. The reflecting colli- mators have been designed with CerVit primaries and minified point sources. The mark lens configuration is comprised of a long focal length lens and wedge manufactured from a single piece of fused silica. The on-line data acquisition and control system will permit real time monitoring of these instrumental param- eters.

1. Introduction

The U.S. Naval Observatory's Automatic Transit Circle (ATC) is presently in its initial stage of performance evaluation of the instrumental constants. The automatic data acquisition for the constants is performed via a control program on an on-line IBM 1800 data acquisition and control system. The observer monitors the procedure from a controlled environment in the southeast corner of the transit circle pavilion. The only task that requires the observer to leave the control room is that of the alignment of the artificial sources in the north and south collimators. However, a method for automating this function is now under consideration.

2. Instrumental Constants

A detailed discussion of the optical system was presented (Klock, 1970) at the IAU 14th General Assembly. It shall suffice to mention here that an artificial source is located in the star tracker off the main optical axis by approximately $31''.7$ in α, and $27''.4$ in δ. This source is used in the conventional method of determining the level and nadir of the instrument with a mercury horizon.

The constants system presently has only one mark lens which is comprised of a long focal length (116 m) lens and wedge ($7°18'$) manufactured from a single piece of Schlieren quality fused silica. A new point source for the mark was designed and developed by H. E. Durgin, T. J. Rafferty, and R. J. Miller of the Observatory staff. The source is observed through a combination BG-18 and GG-400 Schott filter. The requirement for the wedge was generated by the close proximity of the collimator pier to the main instrument. The wedge-lens is mounted below the collimator at the end of the collimator pier closest to the telescope.

The collimator system encompasses two reflecting collimators with 25 cm diameter CerVit primaries and a novel self-contained artificial source. Each source has a range of motion of $\pm 6''$ in both the x and y coordinates. A star of approximately 4th mag. is simulated. The actual diameter of the source is 0.9 mm; however, the optical con- figuration minifies the source by a factor of 100; i.e., the apparent image size is

0.009 mm. As an aid in alignment of one source apparently onto the other, the sources may be blinked alternately at a 15 Hz rate.

3. Data Acquisition and Control Program

As previously mentioned the ATC operates in conjunction with an IBM 1800 process control computer. The system has the capability to send information as well as control signals to the telescope, to receive data and status signals from the telescope, to communicate with the observer via a console typewriter, and to store $1\frac{1}{2}$ million words of data on magnetic disk storage. Results can be printed on a line printer or punched onto cards.

As an aid to system checkout a program was written to allow each individual operation of the ATC to be controlled separately with the results displayed on a television monitor. The data registers can be loaded, read, and displayed in both binary and decimal formats.

The observational control program contains routines to compute the apparent place, to control the complete sequence of an observation, and to communicate with the observer. The latter consists of the issuance of messages concerning errors as well as the progress of the observation and of the receipt of information and commands from the observer. For example, the observer can request that the data be printed on the typewriter simultaneously with its storage on the disk and thus permit him to watch the progress of the observation.

The sequence of an observation begins with the entry of a star or constant number into a keyboard by the observer. He may enter star numbers one at a time or he may enter a list of up to 32 stars and constants. An entire night's observing list can also be stored for use on the disk.

If a star is to be observed, it is found on the disk file containing the 1950 rectangular coordinates. A second file contains the Besselian day numbers and precession matrix. The apparent place is then computed to an accuracy of approximately $0.01''$. The instantaneous refraction is calculated on the basis of the real time readings of the thermometer, barometer, and dew cell (all of which are automatically sampled by the computer). In the case of the constants the setting angle positions of the various light sources are stored. In order to follow the seasonal variations these settings are automatically corrected to the mean of the preceding night's observations.

Simultaneously with the preceding action the telescope is lifted hydraulically 0.05 mm and driven to the apparent declination, and then lowered. The telescope settles into a position within $\pm 15''$ of the desired position. The residual error between this position and the computed position is applied to the declination tracking mirror.

Once a star or constant source is detected and acquired for tracking the declination tracking mirror and right ascension tracking mirror are read out every second for thirty seconds on either side of the transit time for a star, and simply for sixty seconds for an instrumental constant. The computer continually monitors the tracking signals, makes decisions, and issues messages to the observer based on the length of time required to acquire the star and whether its signal is lost at any time (i.e., in the event of a cloud obscuration).

The data are stored on the disk for later reduction. However, certain calculations are performed in real time to allow the computer to check for error conditions. The temperature, pressure, and dew point are continually monitored and checked against similar backup systems. The speed at which certain operations occur is monitored to detect equipment failure. The pressure in the hydraulic lift mechanism is used to determine whether the telescope has settled onto the wyes before the readings are taken.

Mean values for the constants are computed, printed on the console typewriter, and automatically checked against the known range of values. The 1800 system has been used for checking errors in this manner on the six-inch transit circle for five years (Gauss, 1968) and has proven to be invaluable in detecting equipment failure and observer errors.

4. Preliminary Results

The expected accuracy of the ATC can be readily seen through the observational data collected with the constants. The internal repeatability of the tracking mirrors at the present time appears to be better than $0.1''$; for the main declination axis Inductosyn the value is slightly higher, $0.15''$.

References

Gauss, F. S.: 1969, *Bull. Am. Astron. Soc.* **1**, 179.
Klock, B. L.: 1970, *Trans. IAU* **XIVA**, Report Commission 8.

THE RESULTS OF INVESTIGATIONS OF THE PULKOVO HORIZONTAL MERIDIAN CIRCLE (HMC)

G. I. PINIGIN, L. A. SUKHAREV, and G. M. TIMASHKOVA

Pulkovo Observatory, U.S.S.R.

The observations in R.A. with the Pulkovo HMC and investigations of the instrument, reported in *Trans. IAU* **XIVA** (p. 39), are being continued during the past two years (Pinigin, 1972, 1973). Some of the new results are as follows:

(1) The foundation of the HMC was made according to the scheme published by Sukharev (1955). A stability of the north and south collimators is characterized by the azimuth variation of $\pm 0''.4$, which corresponds in practice to the stability of the meridian marks of the Pulkovo large transit instrument.

(2) The mirror of the HMC is a monolithic steel block with a rotation axis. The temperature deformations of it are negligible. The collimation of the first and second reflecting surfaces of the mirror can be represented by the formulas

$$C^I = -0''.29 + 0''.0040\,(t° + 0°.7) - 0''.100\,\Delta T$$
$$C^{II} = -0''.85 - 0''.0045\,(t° + 0°.7) + 0''.060\,\Delta T,$$

where $t°$ is the temperature in the pavilion, $\Delta T = T - 1969.75$ – time in years.

(3) The two coordinate wedge micrometers allow autocollimation measurements to be made by the photoelectric method with an accuracy of $\pm 0''.01 - 0''.02$. This result confirms the good quality of the optical system, which has the following construction:

(a) the objectives of the collimators are composed of two lenses, crown and flint with similar coefficients of linear expansion;

(b) the lenses are put into the self-centering casing of the bearing type. The lenses rest upon the bearings under the action of their weight. The bearings, which are like those of a classical meridian instrument, are fixed on the pillars. The eye-piece systems of the collimators have similar construction. The micrometer tube is fastened in the center of the disc made of crown glass;

(c) the collimator tubes are not connected either with the objectives or the eye-pieces or pillars. They serve for protection from outside influences only. Each tube consists of two elements; a massive steel inner tube and a light aluminium outer one. There is a special ventilating system between them.

(4) The pivots of the HMC mirror rest upon box-wood bearings. Such a system of support preserved the figure of pivots during 1.5 years of operating.

(5) Seasonal variations of the instrumental system (Δn_δ) have not been detected in practice within the limits of $\pm 0°.01$.

(6) The mean error of one determination of right ascension is about $\pm 0°.011 \sec \delta$ for upper culmination and $\pm 0°.013 \sec \delta$ for lower culmination.

All these investigations show that the Pulkovo HMC can be used for precise de-terminations of R.A. At present the HMC is being prepared for observations of declinations.

References

Pinigin, G. I.: 1972, *Trudy 18th Astrometr. Konf.*, p. 158.
Pinigin, G. I.: 1973, *Trudy 19th Astrometr. Konf.*, in press.
Sukharev, L. A.: 1955, *Trudy 12th Astrometr. Konf.*, p. 189.

PROJET DE CENTRE DE RECHERCHES
ASTROMETRIQUES ET GEODYNAMIQUES EN FRANCE

J. KOVALEVSKY

Observatoire de Meudon, France

Résumé. Il y a de nombreux avantages de rassembler, en un même site, un grand choix d'instruments divers en astrométrie et géodynamique, en particulier pour étudier et comprendre les erreurs systématiques de diverses méthodes.

C'est une des raisons pour lesquelles un nouvel observatoire est en train d'être installé près de Grasse.

Le site choisi est le meilleur pour les observations aux astrolabes Danjon. Il entrera en fonction en 1974 et comportera deux astrolabes, un laser pour satellites, un télescope de Schmidt destiné aux observations astrométriques et, plus tard, un laser-lune et d'autres instruments.

Abstract. There are many advantages in having at the same site, a large array of various instruments for astrometry and geodynamics, in particular in order to provide a good insight into the systematic errors of various methods. This is one of the reasons why a new observational center is being equipped now near Grasse. The site chosen is the best found for Danjon astrolabes. It will start its operation in 1974 and will include two astrolabes, a satellite laser, a Doppler satellite receiving station, a Schmidt telescope fit for astrometric observations and, later, a lunar laser and other equipment.

Au cours de ce symposium, on aura discuté de l'apport considérable que l'on peut attendre, dans le domaine de l'astrométrie, des nouvelles méthodes d'observation. Qu'il s'agisse des instruments classiques comme l'instrument méridien ou l'astrolabe, rendus plus performants et automatisés ou qu'il s'agisse de techniques entièrement nouvelles comme le laser-lune, l'interférométrie à longue base ou des satellites spécialisés dans des mesures géodynamiques ou astrométriques, deux problèmes fondamentaux se posent:

(1) Il est nécessaire de prévoir une longue période de recouvrement des observations afin d'assurer la continuité des paramètres observés (mouvement du pôle ou rotation de la terre par exemple); les observations par les deux méthodes doivent être simultanées et être faites dans des conditions aussi voisines que possible, idéalement au même endroit.

(2) Les divers moyens d'observation sont évidemment grevés plus ou moins d'erreurs systématiques. Il est nécessaire d'analyser ces erreurs et, pour cela, comparer les résultats obtenus à partir de techniques différentes. Là encore, les comparaisons se feront dans les meilleures conditions si les instruments sont situés au même lieu.

C'est pourquoi, il apparait qu'il y a un gros avantage, de regrouper dans un observatoire unique l'ensemble des moyens modernes d'observation astrométrique, tout en les faisant voisiner avec des instruments classiques.

Sous l'impulsion d'un certain nombre d'astronomes français, notamment de M. J. Delhaye, actuellement Directeur de l'Institut National d'Astronomie et de Géophysique (INAG), il a été décidé de consacrer une partie des moyens d'équipement vers une rénovation de l'astrométrie. Parallèlement, d'importants moyens spatiaux ont été mis, par le Centre National des Etudes Spatiales (CNES), à la disposition des astronomes et géodésiens français pour poursuivre des études de géodynamique spatiale, relatives surtout au potentiel terrestre, à la géodésie, aux divers mouvements de la terre et de sa croûte, et à la position de la lune.

Ces conditions favorables ont permis de tenter de créer un centre astronomique nouveau, essentiellement consacré à l'astronomie de position et à la dynamique du système terre-lune, en y rassemblant le plus grand nombre possible d'instruments nouveaux pouvant contribuer à ces domaines et en y assurant, par la venue d'équipes suffisamment nombreuses et expérimentées, l'exploitation de ces instruments, leur intercomparaison et l'analyse des résultats d'observations.

C'est en 1965 qu'a commencé la recherche d'un site pour le nouvel observatoire. Il s'agissait alors, dans l'esprit des promoteurs, de trouver un endroit pour placer les instruments d'astronomie de position qui étaient tous situés dans des observatoires englobés dans d'importantes agglomérations urbaines (Paris, Bordeaux, Besançon, etc. ...) et de réaliser en quelque sorte, une opération analogue à celle qui avait été faite pour les instruments d'astrophysique à Saint Michel et au Pic-du-Midi. Il était en effet vite apparu que les qualités du ciel nécessaires pour l'astrophysique (clarté, scintillation faible, faible quantité de vapeur d'eau) étaient très différentes de celles que l'on recherche en astronomie de position (stabilité des images avec des périodes de l'ordre de quelques secondes, pas de réfraction anormale).

La qualité astronomique du ciel avait été testée par des séries d'observation à l'astrolabe. Huit sites ont ainsi été étudiés et comparés par F. Laclare qui a établi un 'rendement' de chaque station proportionnel à la fois à l'erreur probable des observations et au nombre de nuits. Un site s'est avéré être meilleur que les autres. Situé au nord de Grasse, à 50 km de Nice, à une altitude moyenne de 1300 mètres, le plateau de Calern a ainsi été choisi, en 1970, comme site du futur Centre d'Etudes et de Recherches Géodynamiques et Astronomiques (CERGA) (figure 1).

Dès la fin de 1970, une petite station est installée au Calern sous la responsabilité de F. Laclare et des observations régulières d'astrolabe Danjon y sont effectuées. La station horaire est équipée d'une horloge à césium et une station de réception Doppler pour satellites vient d'y être installée. Enfin dans une grotte voisine, deux pendules horizontaux du type Melchior, prêtés par l'Observatoire d'Uccle ont été placés pour des mesures de marées terrestres tandis qu'une campagne de mesures par gravimètre a été effectuée en début d'année.

La qualité du site a été confirmée. Près de 230 nuits par an sont au moins partiellement utilisables pour l'observation sur des périodes continues d'au moins deux heures. L'erreur quadratique moyenne des observations à l'astrolabe est 1,2 fois plus faible qu'à Paris, avec les mêmes observateurs (figure 2).

C'est en été 1974, que le CERGA sera effectivement créé et qu'une trentaine de personnes venant surtout de Paris, viendront en compléter le personnel. Et, en 1974 et 1975, une première tranche d'équipement instrumental sera mise en place. Les instruments suivants y seront alors installés.

1° Un second astrolabe impersonnel Danjon venant de l'observatoire de Paris (?).

2° Le prototype d'un astrolabe photo-électrique avec deux équerres optiques

Fig. 1.   Sites étudiés pour le choix d'un observatoire astronomique en France.

d'angles différents permettant des déterminations absolues de déclinaisons et atteignant la magnitude 7,5.

3° Un laser pour satellites artificiels, conçu pour observer de nuit en poursuite par visée optique ou de jour et de nuit par poursuite automatique, avec une précision de l'ordre de 20 à 30 centimètres sur les mesures de distance.

4° Un téléscope de Schmidt avec un miroir de 150 cm et avec une lame de 90 cm. L'optique a été étudiée de telle façon que le champ soit suffisamment proche d'un plan pour que l'on puisse prendre des clichés de $6° \times 6°$ sur des plaques de verre montées sur un chassis spécial. L'ouverture focale résultante étant de $F/3,5$, on espère avoir une précision meilleure que $0'',1$ sur la position des étoiles. Ce téléscope servira avant tout pour établir un catalogue de position des étoiles, comme première partie d'une nouvelle 'carte de ciel'. Il sera aussi utilisé pour l'observation de certains astres faibles du système solaire (satellites), mais aura plus tard, de plus en plus une vocation astrophysique. Une machine à mesurer automatique est actuellement étudiée pour réduire les clichés de ce téléscope.

5° Un laser-lune dont la précision de tir devrait être initialement de un mètre. Le laser et la partie électronique seront celles qui sont actuellement en place au Pic-du-Midi, mais l'optique, étudiée à l'INAG sera nouvelle. Il s'agira d'un télescope d'environ 150 cm d'ouverture. Cet instrument travaillera de façon continue de jour comme de nuit chaque fois que cela sera possible.

(6) Un ensemble de gravimètre et de pendules horizontaux pour suivre les mouvements du sol de l'observatoire.

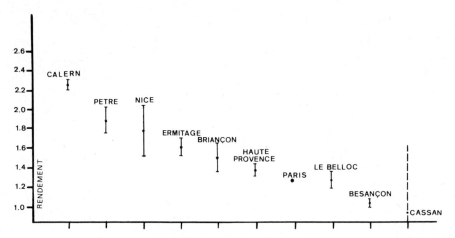

Fig. 2.   Résultats de la prospection astronomique en France. Le rendement d'une station dépend de la qualité des images et du nombre de nuits d'observation par an.

(7) Un terminal relié à un grand calculateur Control Data/7700 du CNES à Toulouse et permettant donc de faire toutes les réductions aussi bien que des travaux théoriques.

Enfin, la proximité de l'observatoire de Nice (50 km) où se trouvent deux lunettes, permettra au personnel du CERGA de poursuivre des études des étoiles doubles et de sélénodésie déjà entreprises. Il est à noter, réciproquement, que les astronomes non résidents, qu'ils soient français ou étrangers, seront les bienvenus pour venir au CERGA faire des recherches dans les domaines des compétences scientifiques de ce centre.

Les projets pour un avenir plus lointain ne manquent pas, mais aucune décision n'a encore été prise à leur sujet. L'effort sera poursuivi pour y mettre des instruments modernes, aussi automatisés que possible, susceptibles de compléter un ensemble destiné à améliorer par tous les moyens possibles la connaissance de la position et des mouvements des astres et contribuer à l'établissement d'un système de référence aussi précis que possible pour toutes les études de cinématique stellaire et de la dynamique du globe terrestre, fondamentale dans la compréhension de sa structure interne.

Mais pour beaucoup des problèmes auxquels nous voulons nous attaquer, notamment dans les domaines de la dynamique de la terre et de la lune et des systèmes de référence, une station aussi complètement équipée soit-elle ne suffit pas. Il serait souhaitable que plusieurs observatoires du type du CERGA soient installés dans le monde, bien distribués géographiquement, qui constitueraient une espèce de base fondamentale servant de référence terrestre aux mesures astrométriques. Les résultats obtenus par diverses techniques seront encore plus facilement comparables et la détection des erreurs systématiques sera encore plus sûre. Pour ne donner qu'un exemple, les différences systématiques constatées entre le mouvement du pôle déduit des observations des satellites TRANSIT peuvent avoir des origines très diverses, l'une étant le fait qu'ils s'appuient sur des observatoires presque tous différents. Une étude de cet effet possible n'est guère envisageable que si plusieurs stations sont simultanément équipées de récepteurs Doppler et d'astrolabes. C'est en ce sens que nous pensons que sera atteinte la plus grande efficacité astronomique des équipements coûteux qui sont mis en place un peu partout. L'expérience que ne manquera pas d'acquérir le CERGA sera, nous le souhaitons, réussie et pourra servir d'exemple dans d'autres régions du monde.

DISCUSSION

Luck: The Division of National Mapping, Australia is installing a Lunar Laser Ranger in the Australian Capital Territory, at an altitude of 1300 m, within 1 km of an ongoing NASA satellite tracking station and within 40 km of the Mount Stromlo PZT, thus satisfying requirements of comparing classical and new techniques – in the southern hemisphere. A laser satellite tracking system might be added in the foreseeable future.

The lunar laser ranger is the 60-in. AFCRL system with CerVit mirror formerly at Mount Lemmon, on indefinite loan from NASA through SAO.

Kovalevsky: I am very happy to see that such a center will exist in Australia and hope that we shall have some co-operative programmes.

Klock: What is your timetable for placing your proposed astrometric center into operation, and will you move the Bordeaux Transit Circle to the new site or construct a new instrument?

Kovalevsky: Most of the instrumentation quoted will be on the site in 1974 or 1975.

The project of a new Transit instrument for CERGA is not yet approved. The idea is to build a new instrument using Requième's techniques tested on the Bordeaux instrument.

Fracastoro: A new 42 in. astrometric telescope has been installed at the Observatory of Turin.

QUE POURRA-T-ON DEDUIRE DES MESURES DE
DISTANCE TERRE–LUNE PAR LASER?

J. KOVALEVSKY

Observatoire de Meudon, France

Résumé. Bien qu'il existe plusieurs stations de télémétrie de la lune par laser, une seule est tout-à-fait opérationnelle : celle de l'Observatoire de McDonald qui donne des erreurs internes d'observation inférieures à 15 cm.

L'interprétation des données fait intervenir un grand nombre de paramètres se rapportant à la terre et à la lune et qui sont données.

Le laser-lune est particulièrement adapté pour déterminer les paramètres relatifs à la lune et, avec les futurs lasers de précision 2 à 3 cm, on peut espérer que cette précision se retrouvera dans la détermination de ces paramètres.

La détermination probable du demi grand axe à 1 cm près en quelques mois, constituerait une nouvelle méthode d'évaluer la partie non conservative du mouvement de la lune. On peut s'attendre à une précision analogue pour la rotation de la lune. La situation pour les paramètres relatifs à la terre (rotation de la terre et mouvement du pôle) n'est pas aussi bonne à cause d'une présentation géométrique plus faible du problème et l'absence d'observations pendant une semaine par mois.

Néanmoins, cela donnera un contrôle externe très utile des résultats obtenus par les méthodes concurrentes (radio-interférométrie, satellites laser et radio).

Abstract. Although several lunar laser ranging stations exist, only one is now fully operational: the McDonald station with internal observational errors of less than 15 cm. The interpretation of the data involves a great number of parameters relative to the Earth and the Moon which are listed.

The lunar laser is particularly fit for those parameters that pertain to the Moon, and with future lasers accurate to 2 or 3 cm, it may be expected that this accuracy will be projected into these parameters. The probable determination of the semi-major axis to 1 cm accuracy for a few months mean would imply a new means of determining the non conservative part of the motion of the Moon. A similar precision is to be expected for the rotation of the Moon. The situation for the Earth parameters (Earth rotation and polar motion) is not so good, because of a rather weak geometry of the problem and the monthly one week gap in the observations. Nevertheless, it will give a very useful external check on other competing methods (radio-interferometry, laser or radio-satellites).

C'est le ler août 1969 que la première distance entre un instrument terrestre et des cataphotes déposés sur la lune a été mesurée à l'observatoire Lick. Depuis, de nombreuses mesures analogues ont été faites, en grande majorité à l'observatoire de McDonald, mais aussi par d'autres équipes américaines, soviétiques, françaises et et japonaises.

Au total trois réflecteurs ont été déposés par les vols Apollo 11, 14 et 15, tandis que deux ensembles de réflecteurs français étaient montés sur les deux Lunakhod soviétiques.

Au cours de ces quatre années, ces instruments ont subi de nombreuses améliorations, tandis que de nouveaux lasers-lune sont en installation, le plus important étant celui que le groupe U.S. Lure va installer à la fin de 1973 aux îles Hawai.

En fait, actuellement, toutes les stations sauf une – celle de l'observatoire de McDonald – en sont à une phase d'expérimentation des techniques. C'est pourquoi, on

Gliese, Murray, and Tucker, 'New Problems in Astrometry', 269–274. All Rights Reserved.

ne peut que se fier aux résultats de cet observatoire pour connaître l'état actuel de la technique en ce domaine.

Le laser est un laser à rubis travaillant à 6943 ångströms avec une énergie de 3 joules, pour un tir toutes les 3 secondes avec une divergence angulaire de 4′. On utilise le télescope de 2,70 m de McDonald. La durée d'une impulsion est d'environ 4 nanosecondes, tandis que la calibration est maintenant faite avec une précision globale de 0,4 ns soit, en distance, 6 cm. Le bruit interne des observations, qui est de 1 ns permet d'avoir actuellement des valeurs moyennes de la distance terre-lune à environ 15 cm prés.

L'interprétation de ces mesures se heurte à de nombreuses difficultés théoriques et pratiques qui sont loin d'être toutes résolues. Si on appelle T et L les centres de masse respectivement de la terre et de la lune, par O la station d'observation et par R le réflecteur sur la lune, l'observation étant faite entre les instants t (instant de tir) et $t + \Delta t$ (instant de réception), et en appelant $t' = t + \Delta t/2$, on a avec une approximation surabondante

$$\mathbf{OR}(t') = -\mathbf{TO}(t') + \mathbf{TL}(t') + \mathbf{LR}(t')$$

et on mesure en fait la distance $D = |\mathbf{OR}(t')|$

En analysant chacun des trois vecteurs composant **OR**, on constate qu'un nombre impressionant de phénomènes entrent dans l'interprétation de la distance D, même une fois corrigée des effets instrumentaux (calibration) ou des effets de propagation (réfraction).

Tous ces vecteurs varient par suite de causes très diverses et, par ailleurs, ils doivent être tous repérés dans un système de référence unique, inertiel, par rapport auquel s'applique la théorie du mouvement de la lune.

1°. *Vecteur* **OT**. Ce vecteur dépend:

– de la position de l'observatoire, donc de ses coordonnées géodésiques par rapport à un système de référence mondial.

– des mouvements de cet observatoire dû aux marées terrestres et, éventuellement, aux dérives continentales.

– de la position de ce système de référence par rapport à un système inertiel, donc des divers paramètres décrivant le mouvement du pôle, la rotation de la terre (TU1) ainsi que la précession et la nutation.

2° *Vecteur* **TL**. Ce vecteur représente le mouvement du centre de gravité de la lune et dépend donc:

– des six paramètres orbitaux ou conditions initiales du mouvement de la lune.

– des paramètres physiques qui déterminent les forces gravitationnelles subies par la lune: masses de la terre et de la lune, éléments de l'orbite de la terre, masses et éléments orbitaux des planètes, grandeur des perturbations des orbites de la terre et des planètes, coefficients principaux du potentiel de la terre.

– des paramètres physiques décrivant l'échange d'énergie dans le système terre-lune due à l'existence de la dissipation par le phénomène de marées et entraînant en particulier un terme quadratique dans la longitude de la lune et une augmentation séculaire du demi-grand axe de son orbite.

3° *Vecteur* **LR**. Ce vecteur dépend :
– de la position du réflecteur dans un système de référence sélénodésique.
– des paramètres de la rotation de la lune qui dépendent eux mêmes des moments principaux d'inertie de la lune ainsi que d'autres moments d'ordre supérieur :
– du paramètre définissant la libration libre de la lune.

De nombreuses études théoriques ont été faites sur ce que l'on pourra effectivement déterminer par cette méthode. En effet, il n'est pas impossible, même s'il y a plusieurs stations terrestres et plusieurs cataphotes sur la lune de déterminer l'ensemble des paramètres dont dépend chaque mesure.

Il ne semble pas qu'il y ait, d'ailleurs, accord général sur ce que l'on peut potentiellement espérer. Une analyse de cette sorte ne peut pas être faite indépendamment d'une analyse parallèle des possibilités offertes par d'autres techniques, compte tenu de leur développement probable et de la couverture spatiale ou temporelle envisagée.

Il y a deux domaines, pour lesquels les laser-lune restent sans concurrents : ce sont ceux qui concernent les vecteurs **TL** et **LR**. En supposant, pour sérier les problèmes, que l'on puisse connaître indépendamment de façon parfaite, le vecteur **OT**, la mesure avec une précision de l'ordre de 15 cm d'une composante de **TL** + **LR** nous

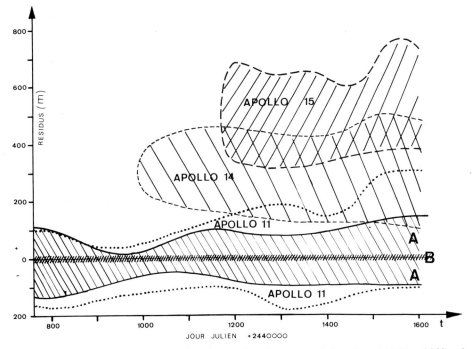

Fig. 1.   Ordre de grandeurs des résidus de réduction des observations de laser-lune à McDonald (d'après Bender *et al.*, 1973). Les régions notées Apollo 11, 14 et 15 indiquent les ordres de grandeur des résidus par rapport à des éphémérides indépendantes des observations. La région A est celle que l'on obtient pour les trois réflecteurs après correction des positions des réflecteurs et de l'observatoire. La région B (< 10 m) indique où se trouvent les résidus après amélioration des paramètres orbitaux.

permet d'atteindre avec une précision comparable les paramètres de rotation de la lune et ceux qui sont relatifs au mouvement orbital.

En ce qui concerne ce dernier, on peut, s'il n'y a pas d'effet systématique inconnu, espérer arriver à définir à 1 cm près le demi-grand axe moyenné sur plusieurs mois, ce qui rendra mesurable l'effet indirect de marées sur la lune estimé à environ 3 centimètres par an (d'après la formule de Spencer Jones). Le gain général par rapport à la connaissance de l'orbite de la lune avant l'ère spatiale (d'après les Improved Lunar Ephemeris) serait donc – en distance – de l'ordre d'un facteur 10^{-4} sur la précision. La figure 1 montre l'amélioration apportée dans la représentation des observations par l'amélioration de certains paramètres (position des réflecteurs et du laser, éléments orbitaux, etc....). L'amélioration des éphémérides correspondantes est d'au moins un ordre de grandeur.

Pour ce type d'études, il ne semble pas qu'il y ait de technique concurrente. Notons cependant que la précision relative avec laquelle on peut déduire la longitude de la lune (relativement à des repères uniquement terrestres) par la méthode laser est plus faible que la distance. Avec 2 stations à 6000 kilomètres et une précision de mesure de 15 cm, on peut obtenir les éléments angulaires du réflecteur à environ 0″.01 près par rapport à une base terrestre. A cette erreur, il faut ajouter les erreurs de passage au système de référence céleste. Ces dernières ne sont éliminées que par les méthodes classiques malheureusement insuffisamment précises de rattachement de la lune aux étoiles (observations d'occultations ou photographie sur fond d'étoiles) et surtout par la radio-interférométrie à longue base entre une source artificielle sur la lune et une radio-source naturelle. C'est cette dernière technique qui, seule, peut concurrencer le laser-lune dans un avenir lointain. Cependant sa mise en oeuvre est extrêmement lourde et difficile; il paraît peu probable que des observations de routine puissent être entreprises par cette technique. Aussi conclurons nous que le laser-lune a d'ores et déjà les capabilités d'accroître, jusqu'à des précisions de l'ordre du centimètre, notre connaissance du mouvement orbital de la lune avec les conséquences géophysiques sur l'évolution du système terre-lune que cela implique.

Le mouvement de rotation de la lune est actuellement mal connu. Les résidus des observations actuelles, au niveau de 5 à 10 mètres, sont essentiellement dus à une mauvaise théorie de la rotation de la lune. Il y a donc ici un gain possible de deux ou trois ordres de grandeur. Il est difficile à dire si un tel gain est suffisant pour atteindre des propriétés physiques de l'intérieur de la lune, mais cela semble peu probable. De toute façon cette connaissance est intrinsèquement utile et la méthode des observations à l'aide de laser lune est la seule qui soit adéquate et rentable. La seule technique concurrente serait celle du Doppler différentiel (improprement appelée radiointerférométrie différentielle). Shapiro a obtenu des résultats préliminaires très prometteurs, mais la lourdeur de la mise en oeuvre de cette technique est telle que la méthode du laser-lune apparaît comme étant meilleure et plus certaine d'être employée de façon suivie.

La question des possibilités offertes par le laser-lune pour déterminer les paramètres de variation du vecteur **OT** est plus délicate à traiter. Les études faites à ce sujet montrent que la précision du laser se répercute de façon assez dégradée sur les éléments relatifs à la rotation de la terre. On arrive, en gros à des précisions journalières de l'ordre de quelques décimètres à partir d'observations à 15 cm près. La difficulté essentielle est que le caractère aléatoire de ces mouvements interdit de cumuler les observations trop longtemps, contrairement à ce qu'on peut faire pour les mouvements relatifs à la lune. Il s'en suit donc que, même s'il y a une dizaine de lasers-lune répartis sur la surface du globe, le pouvoir séparateur spatio-temporel de cette méthode reste limité. Il est à comparer avec les possibilités d'ores et déjà atteintes pour le mouvement du pôle par un système peu précis de poursuite radio des satellites Transit. On atteint maintenant 40 cm pour 2 jours alors que, potentiellement, il y a au moins deux ordres de grandeur à gagner en améliorant la technologie du système. Les préétudes relatives à l'observation de petits satellites sphériques à 3000 kilomètres (LAGEOS) montrent qu'avec des lasers de même qualité que les actuels laser-lune (15 centimètres), on pourra obtenir TU1 et le mouvement du pôle avec beaucoup plus de finesse, simplement parce que, à couverture instrumentale égale, les conditions géométriques sont meilleures. Par ailleurs, l'absence d'observations de la lune pendant la semaine de la nouvelle lune est un grave inconvénient pour des études fines de la rotation de la terre et du mouvement du pôle qui sont nécessaires pour étudier leurs variations brusques, et les corréler avec d'autres phénomènes géophysiques.

En ce qui concerne la détermination des relations entre les systèmes de référence, cette technique ne pouvant pas atteindre les axes stellaires, son efficacité est assez faible, bien qu'il y ait un certain intérêt à relier la position de la lune à des repères terrestres, comme nous l'avons vu plus haut. Mais, de façon générale, les méthodes de radio-interférométrie à longue base sont beaucoup plus efficaces pour ce problème.

Pour nous résumer, nous disons donc que, avec la précision actuelle du laser-lune de l'observatoire de McDonald, le laser-lune permet un fantastique gain en précision pour tous les mouvements de la lune et semble être la technique la plus simple à mettre en oeuvre et la plus efficace. En revanche, pour les paramètres terrestres, elle a des concurrents mieux placés et ne peut prétendre qu'à un rôle d'appoint ou, ce qui est très utile, de contrôle des autres méthodes pour la discussion des erreurs systématiques.

Mais on peut objecter que la précision des laser-lune va augmenter. En effet des lasers donnant 3 centimètres sont en construction. Les limitations des possibilités de cette technique proviennent d'une part des dimensions des cataphotes lunaires dont l'orientation relative à la terre varie avec la libration et de la difficulté de modéliser à mieux que 1 centimètre près l'effet de la réfraction atmosphérique. En estimant à 2 centimètres cette limite, on voit que cela renforce encore nos conclusions sur l'efficacité de la méthode pour l'étude du mouvement de la lune. Mais cela ne modifie pas non plus les conclusions relatives à la terre, car les mêmes améliorations seront apportées aux lasers de télémètre de satellites; d'autre part, comme nous l'avons dit, il y a encore une grande marge de possibilités d'améliorations des systèmes radio-électriques, et enfin, l'interférométrie à longue base aura un grand rôle à jouer dans la recherche du système de référence.

En conclusion, nous dirons que, dans l'avenir, le laser-lune est capable de fournir une précision centimétrique dans l'étude du mouvement de la lune, notamment dans la détermination des termes non gravitationnels et il est compétitif, sans être le meilleur système possible pour les études relatives aux mouvements et aux déformations de la terre.

Bibliographie

Une bibliographie très complète a été donnée par Bender, P. L., Curie, D. G., et al.: 1973, Science **182**, 229.

DISCUSSION

Van Herk: Are the effects of the Earth-tides known so well that they can be corrected for?

Mulholland: The modellability of solid Earth tides depends strongly on location. Near ocean margins, the situation is not good, but a far inland site such as McDonald Observatory behaves very closely in accord with the presently available theory.

Kovalevsky: This is why, for near oceanic sites like those in Europe, it is necessary to monitor the actual motions of the local crust by gravimeters and horizontal pendulums.

Mulholland: It is worth pointing out the significance of a determination of the secular variation of the lunar semi-axis. Kovalevsky's value of 3 cm yr^{-1} assumed a specific value of the secular acceleration in longitude. The conventional value of 11″/cy, determined by Spencer-Jones some 35 years ago, is apparently still unexplainably large for geophysicists. In recent years, four different investigations have resulted in derived values 100% larger than Spencer-Jones' value. Thus, a direct determination of da/dt should be very important.

It is quite true that the lunar problems will dominate the data analysis for some time to come. For example, Williams and others have shown that the higher moments of the lunar figure produce quite sensible effects on the libration, sufficiently so that one may expect that certain 3rd order gravity harmonics will likely soon be best determined from laser ranging.

Murray: Can the vector **OT** be determined adequately from satellite ranging?

Kovalevsky: Not fully. I believe that this will be possible as far as mean position of T and motion of the pole and rotation of the Earth are concerned, using satellites like LAGEOS for instance. The local tidal effects should be measured on the spot with gravimeters and horizontal pendulums.

Murray: Will it ultimately be possible to determine the mean orbit of the moon and hence the ecliptic, relative to the geodetic frame? If so, we could measure the precession from purely distance measurements.

Kovalevsky: I have not given this problem sufficient thought. I believe that it is theoretically possible, but I cannot commit myself on the eventual accuracy of the determination.

SUR LA MESURE DES DISTANCES PAR
TELEMETRIE LASER

R. BOUIGUE

Université de Toulouse, France

Résumé. On fait une analyse des différents procédés de mesure des distances dans le système solaire et dans son voisinage en vue de mettre en évidence les différentes corrections que ces mesures peuvent nécessiter.

Une application particulière faite aux distances obtenues par télémètrie-laser montre que ces corrections ne sont plus actuellement négligeables.

PROSPECTS OF SPACE ASTROMETRY

P. BACCHUS and P. LACROUTE

Observatoire de Strasbourg, France

Abstract. The advantages of making astrometric observations from space are reviewed, and a project for measuring angular separations between stars along great circle arcs of the order of 90°, by means of a telescope mounted on a TD 1 satellite is described. Positions with accuracy of the order of $\pm 0\rlap{.}''01$, parallaxes to $\pm 0\rlap{.}''007$ and proper motions to $\pm 0\rlap{.}''005 \, \mathrm{yr}^{-1}$ for some 40 000 stars brighter than $9\rlap{.}^{\mathrm{m}}5$ would be obtainable in a two year programme.

1. Introduction

The problems in astrometry which may benefit from the specific character of space observations are those related to relative positions of stars, either at small angular separations (double stars) or distributed on the whole sky (reference sphere). The problems of proper motions and of trigonometric parallaxes are of the same nature.

On the other hand, problems dealing with the position of the Earth with respect to the stars (precession, etc.) can benefit only indirectly from space observations.

The advantages of space astrometry over ground based observations are the following:

(1) No atmospheric refraction, and hence no uncertainty affecting its value, and no rapid changes of the refraction

(2) No atmospheric absorption, which renders astrometry in the UV and X-ray wave lengths feasible

(3) No atmospheric diffusion, semi-permanent observations being therefore possible

(4) No gravity, hence no flexure

(5) Better images; stars at smaller angular distances, and fainter stars are within reach.

In Section 3 of this paper, a project submitted to ESRO, implying the launching of a new TD 1 satellite, is described. The order of magnitude of the cost would be of 10^7 F.F., a comparatively small figure, with regard to the results hoped for.

2. Consequences for Astrometry

We assume that relative positions, parallaxes and proper motions of at least 40 000 stars, brighter than $m = 9.5$, with accuracies of $0\rlap{.}''01$, $0\rlap{.}''007$ and $0\rlap{.}''005 \, \mathrm{yr}^{-1}$ respectively will be secured within two years.

This sphere, because of the way it is established, will be 'fitted' to the present fundamental system FK4; although 'fitted' is somewhat ambiguous, because the fit depends on the weight attached to the stars' present position. The weighting system, which should take into account the random errors and the systematic errors, is not easy to set up.

Gliese, Murray, and Tucker, 'New Problems in Astrometry', 277–282. All Rights Reserved.

We shall examine some of the possible uses of this sphere.

(1) The positions and proper motions of the stars observed define a frame of reference. It can be rendered absolute in different ways:

(a) One may use the information on precession constants that are now available

(b) One may use observations of extragalactic objects

(c) One may use celestial mechanics

In any case, the fact that the system is rigidly defined enables us to use simultaneously observations obtained for any region of the sphere, and will yield far more accurate results than those now being used.

(2) The number of stars included in the system will be such that systematic errors in all our catalogues of present positions and of proper motions will easily be removed.

Furthermore, the absolute parallaxes observed without any systematic error for a much larger number of stars than is now available, will enable us to make statistical estimates of luminosities and to study systematic errors in the presently available material.

(3) The reference system obtained will be very well suited for problems of celestial mechanics and for all calculations of plate constants. It is even unnecessarily accurate for this purpose. Its main advantage in this case is that it will be free of systematic errors.

Of course, such a system does not solve all our problems. It will have to be extended to the fainter stars. The bright stars which will not be included in the system because of the presence of neighbouring disturbing stars, will require special observations. This is the case for double stars, which are of interest in themselves. Ground based observations would do, but it would be possible to make space observations, with slightly modified devices, and also a good deal more computing work.

It seems that the availability of such a sphere, which users of our results would find very convenient indeed, demands that we should reconsider all our ground based programs, considering the results obtained by space techniques.

3. A Project for Space Astrometric Observations

3.1. SHORT DESCRIPTION

A project had been submitted in 1967 to the IAU. It has evolved since and was tested by CNES, which could not carry it out, because it could not be modified so as to be launched by a 'Diamant' Rocket.

It appears now that a rather different project based on the use of a TD 1 satellite, could achieve much more easily far better results.

3.2. OPTICS

A complex mirror reflects on a telescope, two fields at right angles. It is most essential that this mirror be rigid, so that the angle between the two fields remains rigorously constant. (The mirror should be made out of a single block.)

The two fields are 40 cm × 15 cm in size. The Cassegrain telescope has a focal length equivalent to 3 m. The constraints of TD 1 impose that the beam be reflected once.

A grid in the focal plane, as shown in Figure 1, enables the measurement of the transit times of the stars.

The grid has a step of 1″; each segment of the figure represents 20 stripes at a

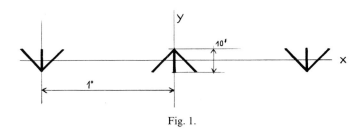

Fig. 1.

distance of 1″. The telescope and the reticule must be rigidly linked together so that the magnification remains as constant as possible (mirrors, reticule and shims would be made of cervit).

3.3. ELECTRONICS

3 photon counters each examine one of the three parts of the grid. The different transit times (one for each part) are used to derive, at the time the star crosses the y axis, the value of y and of the successive derivatives of $x(t)$ and $y(t)$.

The plane containing the two directions observed will be perpendicular to the direction pointing, usually towards the Sun.

The mean crossing time for each part of the grid should be computed on board the satellite. In the course of this work a large proportion of the perturbed data (perturbed by the presence of neighbouring stars) would be eliminated.

Computations show that the crossing times are affected by random errors due to these background stars.

One can also show that the frequency and accuracy of stars at different magnitudes is such that, during each orbital revolution, data would be collected for:

16	stars of	$m<6.5$	$<0″.0010$
35		$m\sim7.5$	$0″.0016$
57		$m\sim8.5$	$0″.0029$
33		$m\sim9.5$	$0″.0052$.
141			

3.4. THE COMPUTATION OF ANGULAR DISTANCES

If two transits occuring at a short time interval correspond to star in the two different fields, then the rotation angle of about 90° defined by the mirror comes in as an unknown experimental constant. In any case, we can observe only separations of stars having transits at small time intervals, and the motion of the satellite must be taken into account in order to derive the x's and y's of the two stars at some given time between the two crossings. Computations show that we can thus evaluate angles between 0° and 4°, and between 86° and 94° without losing accuracy, by taking into account the data collected and the equations of motion of the satellite. The range of angles can be increased somewhat, but in consequence the accuracy diminishes.

It must be understood that these determinations can only be made if the satellite's orientation has not been changed while information was collected. This is an incentive not to change the satellite's orientation too often, but valuable results can already be obtained if this occurs only at intervals of several minutes.

Thus the data collected during one revolution of the satellite will yield with very high accuracy, approximately 6 angular distances per star, or, in other words, 420 arcs on the sphere, half of which between stars at about 90 deg distance.

The great circle arcs connecting two stars belonging to different fields are always determined to an accuracy of better than 0″.01 if the mirror angle is well defined.

Arcs connecting two stars of the same field yield precise information on the value of the arc and on its orientation with respect to the arc that connects one of the two stars to a star in another field, transiting at approximately the same time.

3.5. THE SETTING UP OF A SPHERE

Observations at right angles in a plane perpendicular to the direction of the Sun are most easy and straight forward with the TD 1 satellite, but they cannot suffice to determine completely the reference sphere, since clearly arcs are all measured along a meridian in the ecliptic system of coordinates. Such arcs give information on the latitudes of the stars, but none on their longitudes.

In order to determine its two coordinates, the star must be at the end of at least two arcs intersecting at an angle markedly different from zero or 180°.

We can obtain and measure new series of arcs not on meridians in three different ways:

(1) By changing the orientation of the satellite so that its axis of rotation lies at say 10° from the Sun's direction. These changes could occur at the rate of one every two days.

(2) By displacing the measuring telescope with respect to the satellite so that its axis would make an angle of say 80° (instead of 90°) with the axis of rotation. These changes could be done also at the rate of one in two days; they would require relatively little energy.

(3) By carrying two unorientable measuring telescopes, one at 90°, the other one at 80° from the axis. No displacement would then be necessary, and more information would be collected.

With any of these methods, the majority of the stars observed will be at the intersection of several arcs, carried by two or three great circles.

All these methods can work and yield accurate results; the choice between them will depend, to a large extent, on space technological factors.

Since these arcs are not generally measured at the same epochs; they can be fitted together only by taking into account proper motions and parallaxes. The elimination (or derivation) of these new unknown necessitates three sets of observations, each at intervals of six months. The minimum length of time for the measuring campaign is therefore eighteen months. In fact, a campaign of two years would avoid losing observations due to moonlight.

3.6. REDUCTIONS

The computation of the arcs will be done 'off line'. Stars will be identified, and approximate positions, at a precision of about $0''2$ will be obtained from the Stellar Data Center in Strasbourg.

The corrections for aberration must be carried out. If an accuracy of $0''001$ is required, then the satellite's velocity must be known to 1.5 m s^{-1}.

The instrumental constants must be determined. A crude estimate of the angle between the mirrors can be obtained after each revolution by comparing the 420 arcs that are close to $90°$ and the arcs deduced from the currently known positions. The constants will be thus obtained to about $0''015$.

The position of each star will be improved by using the arcs that link it to other stars and weighting the observations according to the uncertainties of the arcs and of these other stars' positions. For example, if a star has been observed on two orbits, and is thus connected to 12 different stars, its position can be secured to about $0''08$.

At this stage, the scaling factor and the constant of the optical system can be solved for at the same time by comparing the measured arcs with the arcs deduced from the positions mentioned above. With the information deduced from a single revolution, that is 420 arcs, the uncertainty due to the scaling factor can be cut down to $0''004$ for an angular distance of $2°$, and the uncertainty of the angular constant to $0''002$. Thus the stability of the scaling factor and of the angular constant can be constantly kept under control, and their values improved.

We will iterate the computation of the positions and of the constants several times, always taking into account the estimated accuracy of the positions estimated at the step before.

The stars linked to many others by accurately measured angles will have more accurate positions. If the links are less abundant, the accuracy will be poorer; but even for a star observed only once, and linked to another star in the same field, the position will still be obtained to $0''01$.

In fact, our computations indicate an accuracy of $0''003$ for the majority of positions, which would give $0''0015 \text{ yr}^{-1}$ for the proper motions, and $0''002$ for the parallaxes, but we must reckon with time instabilities in the constants, scaling factor and angle between the mirrors; they will have to be known permanently to $0''005$, if the above stated precision is to be achieved. They are perhaps not constant to this accuracy, but they can be measured to better than this, every ten minutes, during the final reduction.

It will be seen that we cannot observe all the stars in this way; double stars and all those perturbed by a star at less than $20''$ separation and $\Delta m < 5$ will be eliminated.

3.7. Advantages of TD 1 for the astrometric project

The limits in the accuracy are set mostly by the stability of the constants and the latter is obviously related to the thermal stability of the optical system. Great care should be taken to avoid the existence of steep thermal gradients, the value of the temperature itself being far less important. TD 1, which retains approximately the same orientation with respect to the perturbating sources, i.e. the Sun and the Earth, is in this respect perfectly well suited to our purpose.

3.8. Accurate locations of X-ray sources

The position of the optical system with respect to the sphere it measures is known at every instant to an accuracy of about $0''.01$. If a device to observe X-ray sources is coupled to the optical system, the location of the X-ray sources will be determined with an accuracy limited only by that of the X-ray device. Accuracy of the order of $1''$ will be easily obtained, which is quite sufficient to identify the optical counterparts beyond doubt.

ASTROMETRY WITH THE LARGE SPACE TELESCOPE

W. F. VAN ALTENA

Yerkes Observatory, Williams Bay, Wis., U.S.A.

O. G. FRANZ

Lowell Observatory, Flagstaff, Ariz., U.S.A.

and

L. W. FREDRICK

Leander McCormick Observatory, Charlottesville, Va., U.S.A.

Abstract. The Large Space Telescope (LST), planned to be launched in 1980, will give an opportunity for astrometric observations of high quality to be secured.

A detailed survey is presented of the potentialities of the LST for studies of proper motions, parallaxes, star clusters, double stars, perturbed motion stars, occultations, fundamental astrometry and solar system observations.

The proposed characteristics and design capabilities of the LST are tabulated.

1. Introduction

In 1980 the National Aeronautics and Space Administration plans to launch into low earth orbit a diffraction limited 3-m telescope for astronomical observations. The Large Space Telescope (LST) will function as an observatory, and as such it will be equipped with several instruments, each designed for specialized observations. Current plans call for return of the LST to Earth by the Space Shuttle for servicing and possible replacement of instruments every two years and for a projected lifetime of 15–20 yr. Due to the very high spatial resolution, large scale and long projected lifetime, astrometry from the LST has been given a very high scientific priority since increases in precision over ground based astrometry by a factor of 10 are anticipated.

To insure that the LST and its instrumentation are designed to solve basic scientific problems, NASA has formed six Instrument Definition Teams composed of scientists and engineers who are charged with the responsibility for developing a preliminary design for the LST instruments. The Astrometry Team consists of W. F. van Altena as Team Leader, O. G. Franz, L. W. Fredrick, and a scientist and two engineers from NASA. NASA has set up a formal communications channel for each team through its liaison NASA scientist so that astronomers wishing to make suggestions to the Team can do so easily. The astrometry liaison scientist is Dr David B. Wood who is located at Goddard Space Flight Center in Greenbelt, Maryland.

The *scientific potential to be gained* through observations from the LST make astrometry a high priority area for all decisions effecting the LST. Depending on the instrumentation finally adopted, it will be possible, for example, to (1) improve the systematic accuracy of the FK4 system in some areas of the sky by a factor of ten or more and tie it into an absolute system of radio galaxies and quasars, (2) mea-

Gliese, Murray, and Tucker, 'New Problems in Astrometry', 283–293. All Rights Reserved.

sure parallaxes and proper motions ten times more accurately than presently possible thus firmly establishing the galactic distance scale and luminosity calibration for the luminous and some peculiar stars, (3) measure angular diameters for stars and the nuclei of galaxies for magnitudes brighter than 11–12, and (4) determine individual masses through 'visual' observations from the LST for about 100 spectroscopic binaries which will enormously improve the mass-luminosity relationship for the very massive stars. These and other areas of potential research are discussed in more detail in the following sections and two tables are given which summarize the basic characteristics of the LST and the capabilities required for the LST if astrometry is to be done from the LST.

2. Parallaxes

Parallax investigations are fundamental to astronomy. All distance determinations ultimately come back to parallaxes determined in the classical sense. In addition, the determination of stellar masses rests heavily upon parallax studies. Therefore, any instrument that promises higher accuracy in parallax determinations should be used and the LST is such an instrument.

LST parallax studies will break down into five areas of study: (1) parallaxes of faint stars (mag. 16 to 19); (2) parallaxes of stars in clusters used for photometric and distance scale calibrations (Hyades, etc.); (3) parallaxes of distant but important objects (planetary nebulae, etc.); (4) parallaxes of sub-dwarfs; and (5) parallaxes of certain key stars for calibration purposes.

Using the best techniques available at the present time, it is possible to determine the relative position of an average well-exposed stellar image on a photographic plate with an accuracy of $\pm 0''.02$ (s.e.). This figure includes errors due to the limited spatial stability of the photographic emulsion, inaccuracies in defining the center of the photographic image, and possibly other instrumental effects. The pointing stability of the LST is projected to be $\pm 0''.005$ rms, therefore a short integration time, or

TABLE I

LST proposed characteristics

Primary mirror		3-m $f/2.2$ CerVit
Telescope length and diameter		12.7 m, 3.7 m
Slew rate		60° in 40 min
Pointing stability		$\pm 0''.005$ (s.e.)
Cameras	Wide field	High resolution
f-ratio	$f/12$	$f/96$
Scale	$5''.7$ mm^{-1}	$0''.7$ mm^{-1}
Field of view	$4'.8$	$36''$
Resolution	$0''.17$	$0''.03$
($\lambda = 3000$ Å)		
Limiting mag.	29	27
($S/N = 2$, 16 min)		
Detectors	50 mm SEC, 20 lp mm^{-1}, 8×10^7 bits	

equivalently, five consecutive position measurements should yield a mean position of $\pm 0''.002$ (s.e.). With an individual relative position determined ten times more precisely than is presently possible from Earth-based telescopes it should then be possible to determine parallaxes ten times more accurately or with an accuracy of $\pm 0''.001$ (s.e.).

This high precision with the LST will yield parallaxes of objects at a distance of 40 pc to 4% while they are now good to only about 20%. Also, parallaxes at distances of 100 pc will be good to 10%. Thus stars used for calibrating photometric indicies, etc., will have much of the accidental scatter reduced, which will result in greatly improved calibrations. Since the Hyades cluster is so important to the cosmic distance scale it should receive high priority.

Many faint stars having large proper motions have been discovered by Giclas, Luyten, and others. These are nearby stars, but many are too faint for current earth based programs. A carefully selected sample of these stars requires parallaxes to add to our knowledge of the total mass in the solar neighborhood and the luminosity of the faintest stars.

Accurate parallaxes for the metal deficient Population II sub-dwarfs will yield their absolute magnitudes which in turn will reveal the distance scale for Population II stars and possibly shed some light on the ratio of hydrogen to helium in the very old stars. The presently available parallaxes are too inaccurate to solve these problems.

Finally, there are unusual objects, for example, the planetary nebulae, for which distances good to 10% will yield a wealth of knowledge and solve some long standing problems. The absolute magnitudes of the hot, degenerate, central stars of planetary nebulae are not known, hence the distances and masses of the nebulae are very uncertain. The confusion is so great that one indirect calibration method puts certain nebulae closer than others while another method has them just the other way around.

To accomplish the measurements of parallaxes we will need at least a five minute of arc field, unless high precision gyroscopes are used in the LST. Constraints upon the telescope and its systems depend upon the instrument used. Measurements at the focal plane equivalent to an accuracy of $0''.003$ are required.

3. Astrometric Studies of Star Groups and Clusters with the LST

The astrometric investigation of groups and clusters of stars with ground based telescopes has, of necessity, been almost completely limited to efforts of deciding which stars in the field of a group or cluster are likely to be members of the aggregate, and to a few investigations of possible overall expansion of such systems.

This limitation has been imposed by the accuracy with which proper motions can be determined from ground-based observations by combining recent measures with those on already existing first-epoch photographic plates obtained at most 60–70 years ago. Moreover, such investigations have been usually limited to stars brighter than the 13th visual mag., a limit dictated by the useful limiting magnitude on early epoch plates. Even if plates were now obtained to much fainter magnitudes, many decades would have to pass until proper motions of an accuracy useful even for membership investigations could be determined.

The standard error of a proper motion determined from one pair of plates with an epoch difference of 50 years is usually at least on the order of $\pm 0''.001$, sufficient, in many instances, for membership studies, but inadequate to study the many interesting and important problems related to internal motions in groups and clusters.

The LST, on the other hand, can be expected to yield, over 5 yr periods, proper motions of stars in clusters and groups not only to magnitudes far beyond the reach of ground based instruments, but with an accuracy that will equal, if not exceed that obtained from ground based observations spanning 50 or more years. If such LST observations can be extended over periods of 10 or even 20 yr, – and most astrometry, even with the LST, will remain long range work – then proper motion determinations with standard errors of perhaps $\pm 0''.0005$ or even better may be expected to be achieved. It is therefore proposed that LST-borne equipment be used for the following proper motion studies:

(1) To determine membership in selected open star clusters to the intrinsic magnitude limit in these clusters. Such a search for faint and the faintest members, in conjunction with appropriate photometric investigations, would not only help to establish the true and complete color-magnitude diagram and the luminosity function of these clusters, but should also result in the detection of the white dwarf members which should be present but, in most instances, have not been found in numbers predicted by evolutionary theory.

(2) To investigate, in clusters of various ages, sizes, and densities, the motions of samples of cluster members of various brightnesses and locations in these clusters, in order to study the character and distribution of their relative (internal) motions. Such studies are of great importance in conjunction with current ideas on the kinematics and the dynamical evolution of star clusters. They should provide answers to such questions as whether or not a Maxwell-Boltzmann velocity distribution has been established, indicating that some stars should have reached escape velocity from the cluster; whether dynamic sorting of stars according to their masses is present; whether distributions of internal motions might reflect star formation at several discrete epochs, etc. The questions are of the utmost importance in conjunction with efforts to understand the formation and the evolution of star clusters, and thus perhaps of stars in general.

4. Objectives of Double Star Research with the LST

Probably the most important purpose of double star research is that of direct determination of stellar masses. The knowledge of stellar masses is basic to our understanding of the internal structure and the evolution of stars; in combination with determinations of absolute bolometric magnitudes, it provides knowledge concerning the empirical mass-luminosity relation, a relationship of fundamental importance

TABLE II

Summary of LST capabilities required for astrometry

Science		Required LST parameters						
		Resol. in Airy Disks (AD)	Field of view	Time resol.	Lim. mag.	Star position measuring accuracy	Integrating time per field	Slew time within field
I. Absolute astrometry	(1)	AD≥0″.05	1″	NA	20th	0″.01	1 min	NA
II. Relative astrometry parallaxes, binaries	(1)	0″.05≤AD<0″.1	1″	NA	<17th	0″.002	5 min	10 min
proper motions	(2)	0″.05≤AD<0″.1	≥5′	NA	<17th	0″.002	5 min	NA
III. Occultations stellar	(1)	AD≥0″.05	1″	100 μs	10th	NA	≤1 s	NA
diameters, double stars, nuclei of galaxies	(2)	AD≥0″.05	1″	100 μs	11–12th	NA	≤1 s	NA

(1) Third Generation Gyros.
(2) Scanning Measurement.
NA Not applicable.

in astrophysics. It is particularly important to push our knowledge of the masses of stars to the limits of the largest and particularly of the smallest stellar masses.

Another major objective of double star research stems from the ever growing evidence that a large number, if not a majority of all stars, exists not as single stars, but as the components of double and multiple systems. This high incidence of duplicity and multiplicity may well be intimately related to the processes of star formation in general. The fact that certain types of objects, such as U Gem variables, metallic-line stars, etc. occur in double stars only, must be of significance. It is particularly intriguing to speculate whether 'single' stars, except perhaps for those ejected from stellar systems as a result of dynamic evolution, may not be those possessing planetary systems.

To increase, if not to complete our information on the true incidence of duplicty and multiplicity among stars and to expand our knowledge on the existence of 'unseen' companions and of components of very low mass is one of the most challenging problems of current double star research. Such an extension of our knowledge of the mass-function to extremely small masses may provide, through inference and extrapolation, important empirical clues concerning the existence and the frequency of other planetary systems.

The components of a binary are almost certainly of common origin, their initial chemical compositions were most likely identical and they have differed possibly only in their original masses. They can therefore provide valuable test cases for current theories of stellar evolution, particularly since their present masses can sometimes be determined or at least reliably estimated in many instances, and provided that their colors and magnitudes can be accurately and individually measured. However, in order to explore fully and exploit this important aspect of double star research, it is necessary to resolve – also photometrically – the very closest pairs and to cover the full range of stellar masses and their combinations from the very largest, usually occuring in spectroscopic binaries optically unresolved and unresolvable by ground based observations, to the very smallest whose existence is mostly inferred from perturbation analyses, but largely unconfirmed and unconfirmable by direct observation from the ground.

For these problems to be resolved or even to be attacked with some assurance of success, it is essential that double star observations be carried out in wavelength regions, magnitude ranges and, most importantly, at angular resolutions beyond those obtainable from the ground with any instrument.

The LST, on the other hand, presents a unique opportunity to carry out observations essential for the satisfactory solution of these problems. It is therefore proposed that LST-borne instrumentation be used to carry out the following programs:

(1) A specific search for all 'unseen' companions whose existence has been inferred from analyses of observed perturbations in the motions of nearby stars. Well over a dozen such perturbations, some with amplitudes of only a few hundredths of an arc second, are now known largely through the work of van de Kamp and his associates and many more will undoubtedly be detected.

The angular separation of the 'unseen' companion from the visible star depends, of course, on the mass ratio between the objects. The masses inferred from perturbation analyses range from about one solar mass to values approaching that of Jupiter's mass.

Mere detection of 'unseen' components at their predicted positions with the LST would, of course, be an outstanding observational accomplishment, since it would provide direct confirmation of the results of determinations and interpretation of stellar perturbations often comparable in amplitude to the errors of observation. Actual measurement of photometric parameters and of a single orbital position of any 'unseen' component, on the other hand, would lead directly to the determination of its mass and luminosity free of the ambiguities often affecting the interpretation of perturbation measurements. Such determinations of the very smallest (stellar) masses would be of the utmost significance. This program should have highest priority among double star projects for the LST.

(2) Of nearly equal importance, however, is a searach for duplicty and multiplicity in nearby stars. The objects to be investigated in this manner may be selected from Gliese's *Catalogue of Nearby Stars*. This search, which should be extended to the infrared to facilitate the detection of faint, cool stellar components, can be expected to contribute significantly to our knowledge on the incidence of binaries and multiples whose components are too faint and too close to be detected by observations from the ground. Once detected, the binaries and multiples, many of which will have rapid orbital motions, should be observed by the LST astrometrically with a frequency and for a duration sufficient to determine their orbits and their masses and thus to provide data for defining the low-mass region of the empirical mass-luminosity relation. It is probable that most of the search for duplicity in the nearby stars can be carried out as a byproduct of occultation studies.

(3) Ground based investigations of spectroscopic binaries contribute to our knowledge on stellar masses only in the case of double-lined spectroscopic binaries which are also eclipsing variables. Observation of non-eclipsing, single-lined spectroscopic binaries with the LST, on the other hand, can be expected to resolve directly many of these objects and permit determination of their 'visual' orbits. Such orbit determinations would, in combination with known spectroscopic orbital elements, yield directly the masses for the components of many spectoscopic binaries and thus contribute importantly to our knowledge on the masses and the mass-luminosity relationship particularly for very massive stars.

(4) A search for duplicity and multiplicity among representative samples of many types of stars other than 'nearby' stars would further enhance our knowledge of the incidence of these phenomena among stars in general. Also, known binaries of separations too close to allow separate photometry of their components from the ground, should be observed with the LST to add to our knowledge on the physical properties of double star components, particularly in the case of pairs of special astrophysical interest.

5. Lunar Occultation Measurements of Stellar Diameters

The knowledge of the diameter of a star is a topic of basic astrophysical interest, and in combination with other data the fundamental physical properties of the star can be determined. Except for the Sun, linear diameters of stars can be determined directly only for eclipsing spectroscopic binaries when the appropriate parameters can be measured. Angular diameters can be measured by several techniques, and if a parallax value is available, a linear diameter can be determined. The methods include the intensity interferometer, the Michelson interferometer, speckle interferometry, and lunar occultations, all of which encounter difficulties from brightness and suitability of available objects. It is difficult to evaluate the development of each of these techniques over the next 10 years, but it seems likely that improvements in the interferometric techniques may provide a considerable increase in the number of observed angular diameters. Nevertheless, lunar occultation results may be the best way of obtaining stellar diameters, at least for later-type stars.

Occultation observations could be useful for double star work in at least two ways. Although a single observation yields only a projected separation and magnitude difference, this information will be sufficient for a study of the frequency of duplicity in stars. Thus, as a byproduct of the occultation diameter measurements, we get statistics on the frequency of occurrence of double and multiple star systems. Also, occultation observations of that class of objects for which duplicity has been invoked to explain their unusual behavior would be a crucial test of that hypothesis. In addition, it may also be possible to measure the angular diameters of the nuclei of Seyfert or N-type galaxies in some cases through lunar occultations.

The principal advantage of spacecraft observation of a lunar occultation is the freedom from scintillation noise and atmospherically scattered moonlight, which constitute the major sources of noise in Earth-bound observations. The non-atmospheric component of the noise is the stochastic noise due to the number of detected photons in the sampling interval, but the characteristics of stochastic noise are well understood and predictable. Other advantages are the lack of weather in space (which reduces the number of Earth-based observations by a factor of three) and the larger number of events which can be observed (due to the larger area covered and uncovered by the Moon as seen from a fast-moving satellite). In addition, the lunar motion across the star field as seen from the satellite is complex, and multiple occultations are possible near the altitude where the apparent lunar motion becomes very slow and finally is retrograde. Multiple observations at different contact angles could remove ambiguities caused by irregularities in the lunar surface. In cases where only one occultation is seen, observations in two or more wavelength regions can also resolve ambiguities caused by lunar limb distortions. Supplementary Earth-based observations would also be of value in this respect (in cases where the occultation could be seen on earth). Ultraviolet observations, which are most useful for early-type stars, should be possible, but due to the dependence of the fringe pattern on wavelength, smaller rocks (15 m at 1500 Å) could cause trouble (whereas the effective region of the lunar limb contributing to the occultation event for a point source at 6000 Å is about 30 m). However, spatial resolution is greater for shorter wavelengths.

The telescope velocity of the LST will be about 14 times greater than that of an Earth-based telescope at the point of central contact when the Moon's apparent velocity is greatest relative to the spacecraft. Due to this greater velocity, we will have proportionately more opportunities to observe occultations. The lunar motion at this point is about $3'' \text{ s}^{-1}$, and the largest acceptable response time to record the fringe pattern is about 200 μs per reading. In many cases, of course, the apparent lunar motion will be slower, and in a few special cases (near the retrograde motion) the occultation may last longer than is seen on Earth.

The time resolution required for lunar occultation studies could also be used for pulsar studies. In other words, the astrometric instrument to be used for occultations could probably also serve as a high speed photometer.

While the telescope aperture (3 m for the LST) imposes a lower limit on the detectable diameter of a star of about $0''.0015$ (due to destructive interference of fringes from opposite points on the mirror), this is not a serious problem. A 500 Å optical bandwidth imposes a similar limit to the resulting measurement. It may be possible to place a mask of, say, 1 m in the perpendicular direction (to the path of the star across the lunar limb) across the center of the telescope aperture to reduce the fringe interference problem and thus allow smaller angular diameters to be detected. If the star is bright enough so that a narrow or intermediate band filter can be used, an additional resolution factor may be gained.

The problems caused by irregularities in the lunar limb will remain even with simultaneous observations in different wavelengths, multiple occultations when circumstances permit, and supplementary Earth-based observations. However, considering the relatively small amount of time required to observe an occultation, it seems that measurements of angular diameters of stars with the LST will provide much useful data with little additional effort in terms of instrumentation and design characteristics.

6. Fundamental Astrometry from the LST

Improvements in the fundamental system FK4 are possible with the LST and should be given high priority in the observing program. At the present time, our knowledge of Oort's constant B of galactic rotation and of the distances of intermediate luminosity stars, through the use of secular parallaxes, is dependent on the FK4 being an inertial reference system. Systematic errors in the FK4 result in similar errors in the derived quantities which can seriously affect our understanding of the structure and dynamics of the Galaxy. The FK4 and systems based upon it, such as the AGK3 and the Smithsonian Star Catalogue, are used for the determination of positions of solar system objects whose calculated orbits may be in error due to systematic errors in the FK4. These errors can seriously hamper projected observations of these objects from fly-by spacecraft or their later recovery after many years from Earth-based telescopes. Radio astrometry has progressed to a point where an intensive effort should be made to tie the radio absolute positions into the FK4 system, or vice versa, since the radio galaxies and quasars will certainly provide the best inertial reference system available. The LST has the potential to accurately accomplish this task because its large aperture will permit the observation of galaxies to the 20th mag. while also allowing observations of the FK4 stars.

The potential of the LST for fundamental astrometry depends on the use of new high precision gyroscopes such as the Third Generation Gyros (TGG). The mean error of the FK4 system is given as $\pm 0\rlap{.}''014$ to $0\rlap{.}''040$ in declination and $\pm 0\rlap{.}''015$ to $0\rlap{.}''135$ in right ascension, where the highest systematic accuracy is obtained near the equator and the lowest near the south pole. However, recent fundamental observations indicate that the errors may be several times larger in the southern hemisphere than estimated in the FK4. The projected accuracy of the TGG is $\pm 0\rlap{.}''04$ (m.e.) over angles of $10°$ and $\pm 0\rlap{.}''15$ over angles of $60°$ for a single measurement with a potential limiting accuracy approximately ten times better through repeated measurement. It seems likely that through repeated measurement and through the use of a network of observations over the sky it will be possible to achieve a significant improvement over the present fundamental system, perhaps by a factor of 10 or more in some areas of the sky.

It may be possible to obtain most of the fundamental observations needed for improvement of the FK4 through the necessary asquisition of guide stars for spectroscopic, photometric, etc. observations. The careful, rather than random, selection of guide stars will yield simultaneously highly accurate relative positions for the two guide stars and the object being measured and also the accurate inertial positions of those three stars.

7. Solar System Observations

In general, Earth-based solar system observations are hampered by insufficient resolution and limited by seeing conditions and scattered light. The LST provides an immediate solution to many problems arising from insufficient resolution. Currently, for instance, the diameter of Pluto is poorly known. Greatly improved observations of the diameters of asteroids and of Pluto could be made with the LST. The mirror coatings in the LST should be smooth enough so that scattered light will not be a serious problem. Low scattered light levels would allow observations to determine, for example, the outward extent and detailed structure of the rings of Saturn.

The LST could also be useful in following interesting comets to large heliocentric distances. The tracking could be accomplished by taking observations of the comet on two successive satellite orbits and digitally subtracting them on board to determine the comets motion.

DISCUSSION

Mulholland: The SEC vidicon used in Mariner-type spacecraft is quite inadequate for astrometric purposes, due to instabilities and distortions. One would hope that one of the newer but much better imaging systems would be used. The alleged reliability of the vidicon is no advantage if the data are useless.

The utility of LST for lunar occultations may be marginal compared to smaller Earth-bound telescopes, because the time of an event will be sensibly different from one side of the mirror to the other. This will cause the interference fringes to be smeared in a way that may degrade the information content, particularly in the determination of stellar diameters.

Moffet: Why should you use the LST for occultation observations? These can be done just as well and infinitely more cheaply from the Earth's surface. Use the LST only for those things which cannot possibly be done from the Earth's surface.

Van Altena: Earth based occultation studies are limited by the noise introduced by the atmosphere. Gains can be made by observing from the LST. It should be kept in mind that an occultation requires only about one second of observing time to obtain very significant and valuable stellar angular diameters, and information on duplicity and multiplicity of stars.

Kovalevsky: Could the author explain how he intends to improve the FK4 system as it is now by a factor of 5 to 10 with angular observations good to $0''1$?

Van Altena: The potential pointing accuracy is one part in 10^6 of the angle traversed, which yields an accuracy of about $\pm 0''04$ for a $10°$ angle and $\pm 0''15$ for a $60°$ angle. Repeated measurements should reduce the error to a value substantially below the systematic errors present in the FK4 which are as large as $0''8$ in the southern hemisphere.

Eichhorn: There is another technique for determining accurately large angles between stars, possibly with an accuracy corresponding to a standard error in the order of $0''1$ or so, which was investigated by Carol Williams at Tampa. It consists in topocentric observations of artificial satellites against a stellar background on photographic plates. There is a great deal of material already available to carry out this project in the satellite plates which have been taken so far.

This is not to say that the LST should not be used for the improvement of the absolute angular distances between the stars, especially since this is – as Strand pointed out – an unavoidable by-product of all the other work carried out on it.

Wood: Relevant to the valuable suggested use of the LST for obtaining data on binary stars it may be worth mentioning that the University of Sydney (Prof. Hanbury Brown) is at present seeking funds to build a large intensity interferometer which will be devoted partly to observation of binary stars to obtain accurate parallaxes including parallaxes.

Vasilevskis: Since the LST will not be an astrometric telescope, its share for astrometry is expected to be limited. Therefore, priorities will need to be assigned to various programmes. I think that the highest priority should go to close binaries.

Strand: I believe the programme of improving star positions with the gyros is the only astrometric programme which might have a chance to be included in the LST, because of conflict with many far more sophisticated astrophysical programmes.

Fricke: Although the positional accuracy which can be reached with the space telescope is not exceedingly high compared with modern fundamental methods, the advantages are considerable for faint stars.

UPGRADING OF THE LICK-GAERTNER AUTOMATIC
MEASURING SYSTEM

L. B. ROBINSON and S. VASILEVSKIS

Lick Observatory, Santa Cruz, Calif., U.S.A.

Abstract. The Automatic Measuring System (*Lick Obs. Bull.*, No. 598, 1971) was built according to a design developed in 1959. Technological advances since then, particularly in small computers and in electronic component reliability, offered an opportunity for a substantial upgrading of the system. Most of the electronic circuitry has been replaced using modern components, and a PDP-8 computer has been added to the system. Punched card equipment has been replaced by magnetic tapes.

The Lick Automatic Measuring System was built by the Gaertner Scientific Corporation, Chicago, in accordance with a design formulated in 1959, and it was installed at the Santa Cruz laboratories of Lick Observatory in 1967. This paper describes work done recently to upgrade the performance of the machine; not so much to increase accuracy of measurement, which has always been satisfactory, but to improve the convenience of operation, increase the rate at which measurements can be done, and to reduce the amount of human supervision required for its normal operation.

The System consists of two components: the Survey Machine and the Automatic Measuring Engine. Since details of construction and past operation are given elsewhere (*Lick Obs. Bull.*, No. 598. 1971), only a general outline will be offered here.

The Survey Machine is used for initial inspection of plates and selection of objects for subsequent measurement. Two plates up to 17×17 in. each can be surveyed simultaneously by means of projection on a screen. Originally, approximate coordinates of selected objects were measured and recorded on punched cards in this process of surveying. These cards then served as input for the automatic measuring process. A card was read, the Automatic Engine located the object, and then photometric and precise coordinate measurements were made and punched on a card. Normally this process was automatically repeated until the last card of the survey deck was read.

The Automatic Measuring System has been heavily used, with a remarkable increase in precision and efficiency over traditional methods. There have occurred, however, many interruptions in the automatic process, caused by occasionally insufficient accuracy of input coordinates, by errors and failures in the card reader and punch, in the relays which controlled the measuring sequence, and by occasional mechanical malfunctions. The number of interruptions has been large enough that unattended operation of the machine was not feasible. The attendant had such tasks as manual centering of poorly located stars, clearing the card punch or reader in case of a malfunction, and restarting the operation in cases when no star was found, or when the servo system could not center an image for some reason. These interruptions were frequent enough to cause us to consider an improvement of the machine by utilizing technological advances that have occurred since it was designed and built.

Originally we considered merely replacing the most troublesome relay circuitry by

Gliese, Murray, and Tucker, 'New Problems in Astrometry', 295–298. All Rights Reserved.

solid state electronics and adding more electronic controls to reduce the need for operator supervision. However, modifications to a special purpose control system often produce unintended results that require more modifications. Such patchwork changes are expensive in engineering time and are also liable to cause difficulties in maintenance. Furthermore, future changes and improvements would be equally expensive and difficult. On the other hand, the use of an on-line computer as a controller means that future improvements in the control logic and in the operating procedure can be made and tested by the user, without the need of engineering support, and with full assurance that a change which proves undesirable can be eliminated by just going back to a previous program. It is also true that a computer, because it is a mass-produced and thoroughly tested instrument, should give much better reliability than could be expected of any one-of-kind special control circuitry. Consequently, the decision was made to use a small computer to provide the additional control needed for the Automatic Measuring System.

During the past winter, the control system for the Measuring Engine has been completely redesigned and rebuilt. The work has been done in two stages. We first replaced the relay controls by solid state circuitry, thus eliminating a major source of interruptions and trouble calls. The machine was then operated normally for several months to test the new components. In June we removed the card reading and punch equipment, and associated electronics, replacing these with a PDP-8 computer, plus magnetic tapes. The computer was also connected to the new solid-state circuitry so that it can monitor and supervise each step of the measuring process.

The second stage of the upgrading project is almost complete. The computer can now automatically correct for plate misalignment, rapidly locate stars on the plate and control the measuring process. A primitive data acquisition program is in operation. However, we have not yet completed the rather large programming job to operate the system efficiently. Also, detailed tests of measurement precision with the new system have yet to be carried out.

The third stage of upgrading will be to connect the manual Survey Machine to the computer instead of the present card punch. This will require only minor hardware additions. Some careful programming will be required to allow the survey and measuring operation to use the computer simultaneously without wasting time for either of these operations.

When the upgrading is done, the PDP-8 will control all data acquisition, data formatting and storage on magnetic tape. Reduction of final output data will continue to be carried out at the campus computer center using existing programs, with minor changes because the input is now from magnetic tape rather than punched cards.

It is well known that programming a small computer for on-line work can be a very big job. We believe, however, that it will be possible to prepare the necessary programs to operate the Automatic Measuring System without an undue expenditure of programming time. This is partly because Lick Observatory has already installed two other identical PDP-8 computers for other applications, and much of the existing proven software can be used in this new project.

The FOCAL programming language (developed by Digital Equipment Corporation) has been chosen for this work; it is very similar to the BASIC language and is easily mastered. The floating point arithmetic with 10 digit precision is rather slow, but more than fast enough for our needs, where the main time delay occurs while waiting for mechanical motions in the Measuring Engine. At Lick Observatory, we have added a large number of additional commands, so that hardware devices such as the Automatic Measuring System are controlled by simple program statements which are easily learned and used. Program sequences using the commands are prepared by typing on a teletype and may be immediately tested, used and stored.

Programs are stored on a miniature magnetic tape (DECTAPE) which is fully addressable and on which individual segments of information can be written without disturbing other data. Programs are written, stored on tape, recalled, edited and stored again as often as desired. No compiling is necessary because the language is an interpretive one. Programs can link together, one program calling another as a subroutine from tape. A program is typically equivalent in computing power to about a page of FORTRAN coding. Over a hundred such programs can be linked together on one magnetic tape, so that even with our small computer we can prepare and use very large and complex program sequences.

It is expected that the upgrading will be completed before the end of this year (1973), with a resulting increase in precision, reliability and efficiency. While we hesitate to speculate on the increase of precision, before thorough tests are carried out, the reliability and efficiency can be discussed from the experience and from the design, based on known facts about operation with on-line computers.

As mentioned, the troublesome relays have been replaced by tested and reliable solid-state circuitry. Furthermore, the original hard-wired unit for control of machine operations, and the mechanical readers and punches with associated electronics have been replaced by the PDP-8 computer, again a tested and reliable piece of equipment. These changes should greatly reduce the frequency of service calls to take care of malfunctions. In case of a malfunction, however, the computer should permit easy diagnosis of the cause.

The increase in reliability will be one but not the only reason for increased efficiency. As mentioned at the outset, two plates of the same field can be surveyed simultaneously. Only one plate, located in the fixed plate adapter whose coordinate reference corresponds to that of the Automatic Measuring Engine, could be measured automatically without further adjustments. For automatic measurement of the other plate either a mechanical alignment had to be made or a new input deck had to be produced by the campus computer from initial manual measurement of three or more control stars. The need of adjustments was especially time consuming in case many plates of the same field had to be measured, as in parallax work. Depending on number of plates or of objects to be measured, either mechanical or off-line computer adjustments were made, or the plate motion was even controlled manually when the number of objects was not large enough to justify these alignments. Using the on-line computer, of course, the mathematical adjustment of the input coordinates will be fully automatic after the operator initially identifies a few control stars.

A perfect alignment of the plate on the Measuring Engine did not guarantee automatic measurement of every object. Occasionally insufficient care in centering while surveying, or large proper motion may lead to positioning the object completely outside the area of the spinning-sector scanner. In the past this would interrupt the automatic measuring sequence. Now the effect of approximately known proper motion can be taken care of by the computer. If the object is not found for other reasons, an automatic search will be carried out, and this fact will be recorded on the teletype. Regardless of finding or not finding the object, the automatic operation will not stop. When one or more subsequent plates of the same field are measured, the photometric and coordinate measurements will be compared with those of the previous plate, and if the discrepancy exceeds a chosen limit, the fact will again be recorded by the teletype, without interrupting the automatic sequence. Thus the operator can leave the Automatic Engine alone after starting it, and then return only to inspect the causes of typed messages. Before the plate is removed, suspicious cases can be rechecked while the operator watches the operation with a closed circuit television monitor. The computer program could also remeasure suspicious cases before the operator returns. The typed messages and automatic error checking should greatly reduce the editing work which in the past was a time-consuming task in detecting mispunches and spurious measurements.

Finally, even mechanical motions can be made more efficient. In the past there was one slewing and two setting speeds for the heavy plate adapter. In order to prevent unacceptable jolts in switching from one to another speed, the range of speeds had to be quite limited, but now the computer can cause continuous acceleration and deceleration.

The above mentioned reasons for increased efficiency are those planned at present. The versatility of computer control and data processing may offer other improvements not visualized at present but which may become obvious and feasible with experience gained.

Acknowledgement

The project of upgrading is being supported by the National Science Foundation, Grant GP-32459.

DISCUSSION

Wall: Does the machine introduce a magnitude equation, and if so is it removed by the computer programme?

Vasilevskis: Yes, a magnitude equation is introduced, and it is not removed yet by the computer. We may do it in future, however. In our astronomical programme the magnitude effect is nearly eliminated, because

(1) We are interested in differential measurements of nearly identical plates, and

(2) An objective grating is used, and grating images (1st or 2nd order spectra) are measured for bright stars.

THE COSMOS PLATE MEASURING MACHINE

N. M. PRATT

Royal Observatory, Edinburgh, U.K.

1. Introduction

The new 48-in. Schmidt telescope of the U.K. Science Research Council is being commissioned at Siding Spring in Australia. Its potential both for extra-galactic and galactic research is enormous but this potential can only be fully exploited with an automatic measuring machine.

Using the prototype GALAXY machine as a basis, discussions were begun with FCS Controls Ltd, of Haddington, Scotland, in 1969 and have resulted in a new machine which is scheduled to have its acceptance trials in September 1973.

2. The New Machine

The only parts of GALAXY which have been retained in the new machine are the plate carriage, with its associated hydraulic and Moiré fringe position systems, and the two basic scanning systems. The special purpose electronics, including the servo systems, have been redesigned and a Honeywell 316 computer has been introduced to supervise the operation of the machine.

The raster scanning system provides the means for a rapid survey of plates for galaxies, stars and nebulae with approximate measures; it can also be used in a transmission mapping mode for high resolution studies of galaxies and nebulae. The spiral scanning system has been modified so that it can make precise measurements of position and size of both galaxies and stars. This has been achieved by generalising the scanning pattern from circles to ellipses whose eccentricity is automatically matched to that of the object being measured. This system can also be adapted to the measurement of objective prism spectra.

The plate carriage can accommodate plates up to 350 mm sq and all output is to magnetic tape.

3. Coarse Measurement Mode

The spot size of the raster scan on the plate can be either 8, 16 or 32 μm, the width of the raster lane being 128 times the spot diameter. The output for each image detected is:

X and Y coordinates;
the area of the image;
the X extent;
the Y extent;

Gliese, Murray, and Tucker, 'New Problems in Astrometry', 299–300. All Rights Reserved.
Copyright © 1974 by the IAU.

the minimum transmission in terms of a greyness scale with approximately 90 levels;

the quadrant of orientation.

This information may be sufficient to separate stars from galaxies and other objects. The positions of 'good' images are accurate to a fraction of the spot size.

The estimated rate of scanning with the 8 μm spot is 2 mm^2 s^{-1}, or one plate 350×350 mm in 17 hours. With the 16 μm spot the rate of scan is four times faster, but with lower resolution. The data for approximately 10^6 images can be stored on one magnetic tape.

4. Mapping Mode

The raster scan can also be used in a digital transmission mapping mode. In this mode the transmission through each spot size element is integrated and digitised in terms of the greyness scale. Scanning is at the same rate as in coarse measurement mode. With the 8 μm spot, the data from approximately 1500 mm^2 of plate can be stored on one magnetic tape.

5. Fine Measurement Mode

The direct computer control permits greater versatility in the profile matching system. In addition to the normal all image technique, it is possible to measure profiles in narrow transmission ranges thus giving a contour mapping facility in the form of best fitting ellipses at specified transmission levels. It is also possible to measure the transmission in an annulus outside an image.

The bit size of the X and Y coordinates has been reduced from 1 μm to 0.5 μm. Three interchangeable optical systems are available enabling different ranges of image size to be measured.

6. Spectrum Scanning Mode

For measuring spectra, the spiral scan is given the form of an elongated ellipse. The extremities of the major axis are used to centre on the spectrum while the minor axis is used to measure either the transmission at intervals along the spectrum, or the positions and strengths of spectral lines.

DISCUSSION

Klock: Is the figure you quoted of $\pm 0.5 \mu$ for position to be interpreted as repeatability?

Murray: It is the least count of read out.

Klock: What is the resolution limit of the machine, that is, how close may two images be and still be successfully measured?

Murray: Stars with a clear separation of at least one scan increment will certainly be resolved.

A NEW HIGH SPEED MEASURING MACHINE FOR ASTROMETRIC PLATES

P. M. ROUTLY

U.S. Naval Observatory, Washington, D.C., U.S.A.

Abstract. A new 'Starscan' measuring machine has been built by Optronics International, Chelmsford, Mass., to specifications drawn up by the U.S. Naval Observatory. The accuracy and speed of operation of Starscan have been optimized for the measurement of astrometric plates. The following features in the design and operation of Starscan are discussed: (a) the granite stages, (b) interferometric stage position encoders, (c) the star centroid and integrated opacity detector, (d) the image profile display system, and (e) the function of the computer in machine control and data manipulation. The performance of Starscan, in terms of operating speed and accuracy, is reported for astrometric plates.

Gliese, Murray, and Tucker, 'New Problems in Astrometry', 301. All Rights Reserved.
Copyright © 1974 by the IAU.

THE INFLUENCE OF THE REDUCTION MODEL ON THE SYSTEMATIC ACCURACY OF ADJUSTED PARAMETERS

H. EICHHORN and A. E. RUST

University of South Florida, Tampa, Fla., U.S.A.

Abstract. Everybody who deals with the numerical determination of parameters from observations is familiar with the phenomenon that the 'external' errors of these parameters are invariably larger than their 'internal' ones. This phenomenon is traditionally explained by the presence of systematic errors.

Since it is virtually impossible to use a completely rigorous reduction model in any astrometric problem, an 'interpolation' model must normally be employed in which there are fewer parameters (unknowns) than in the rigorous one. It is shown that under these circumstances, the rms errors of the adjustment parameters are always underestimated, which creates the illusion that the results computed with these parameters are more accurate than they actually are.

STAR POSITIONS FROM OVERLAPPING PLATES IN THE CAPE ZONE −40° TO −52°

S. V. M. CLUBE

Royal Observatory, Edinburgh, U.K.

Abstract. A modification of the normal regression models used for the determination of star positions is described. Some preliminary results relating to accuracy are given.

I would like to make some brief comments in relation to the systematic errors that may be introduced to photographic star positions as a result of incorrect modelling. The changes in procedure described are being applied to the determination of star positions from overlapping plates of the recent Cape survey, but the results are at present provisional.

Consider first the 'simple' problem of estimating m in the functional relationship $y = mx$ between an independent variable x and a dependent variable y. It is not widely appreciated that the least squares procedure by which m is found differs according to the origin of the residuals $r = y - mx$. Three situations may be distinguished (sampling errors in m are neglected):

(1) The functional relationship is exact (i.e. the model is correct) and there is no error in x, by which it follows that all the error occurs in y: in this case, $\sigma_{rr} = \sigma_{yy}$ and the 'standard' least squares procedure applies.

(2) The functional relationship is exact, but there is error in both x and y, which may or may not be correlated: in this case, $\sigma_{rr} = \sigma_{yy} + 2m\sigma_{xy} + m^2\sigma_{xx}$. There is an extensive literature concerning this case in particular and in more general circumstances, which is not discussed any further here. Case (1) is obviously a special example of case (2).

(3) The functional relationship is not exact (i.e. the model is only approximate), and x and y are both error free. In this case, the adopted relationship is arbitrary but is presumably guided by some physics which is assumed relevant to the situation. The variance σ_{rr} can be written as $x^2\sigma_{mm}$ assuming a normal distribution for the m values though this may be as arbitrary as the adopted functional relationship.

The third case seems to have been identified clearly only very recently in the statistical literature (Fisk, 1967) and is known as a 'regression model of the second kind', but the most general case would be a combination of cases (2) and (3). An example of the most general case arises in the study of stellar kinematics and an appropriate least squares technique of analysis has been developed elsewhere (Clube, 1972). The derivation of plate constants confronts us with a very similar problem. In determining an improved standard coordinate ξ from photographic measures (x, y), we may write each residual

$$r = \xi - (ax + by + c)$$

Gliese, Murray, and Tucker, 'New Problems in Astrometry', 305–306. All Rights Reserved.

and commonly analyse for the plate constants (a, b, c) as if all the residual error originates in ξ, x, y whereas in practice there is likely to be a considerable contribution to r arising from shortcomings in the model (i.e. either too few or too many or wrong terms). A more satisfactory representation of σ_{rr} should reflect the likely deterioration in the model as $(x^2 + y^2)^{1/2}$ increases, and we could for example write

$$\sigma_{rr} = \sigma_{\xi\xi} + a^2 \sigma_{xx} + b^2 \sigma_{yy} + \beta^2 (x^2 + y^2) \equiv \alpha^2 + \beta^2 (x^2 + y^2).$$

This then introduces two new 'plate constants' α, β and necessitate an adjustment of the weights of the condition equations contributing to the estimation of the plate constants.

In applying this procedure to the Cape survey plates, it is found initially that $\alpha/\beta \gg x$, y because the dominant source of uncertainty is due to the meridian circle positions, but subsequently using averaged positions from overlapping plates, $\alpha/\beta \sim$ one half the plate width revealing that modelling errors towards the plate edge are comparable to the photographic errors. As a result, it now becomes possible to introduce an effective criterion by which adjustments to the model may be judged; improvements are those which cause $\beta \to 0$. It is hoped that the joint application of this procedure with the 'overlapping' constraints will result in some control over any systematic errors introduced photographically.

References

Clube, S. V. M.: 1972, *Quart. J. Roy. Astron. Soc.* **13**, 185.
Fisk, P. R.: 1967, *J. Roy. Statist. Soc.* **B29**, 266.

DISCUSSION

Eichhorn: It is very encouraging to see that the method of least squares is beginning to be regarded more critically than used to be the case. There have, in the past, been a lot of sins committed against the spirit of Gauss by applying least squares adjustments in ways that they were not meant to be used.

The problem of more than one observation in any equation of condition has, by the way, been solved by Denning (*Statistical Adjustment of Data*, N.Y. 1942) and extended, in a very general and elegant form, to the case of correlated observations by D. Brown in a paper 'A matrix Treatment of the General Case of Least Squares', published 1955 as Report No. 937 by the Ballistics Research Laboratory at Aberdeen Proving Grounds, Aberdeen, Maryland, U. S.A.

Clube: These papers are certainly elegant and relevant to the plate constants problem, but do not cover the equally serious problem of uncertainties in the physical model.

A GENERAL COMPUTER PROGRAM FOR THE APPLICATION OF THE RIGOROUS BLOCK-ADJUSTMENT SOLUTION IN PHOTOGRAPHIC ASTROMETRY*

CHR. DE VEGT

Hamburger Sternwarte, F.R. Germany

and

H. EBNER

Universität Stuttgart, Institut für Photogrammetrie, F.R. Germany

Abstract. The evaluation of stellar positions from overlapping plates by rigorous block-adjustment methods leads to a very large system of normal equations containing both stellar positions and plate constants as unknowns. These general normal equations can be reduced to a subsystem with banded-bordered coefficient matrix, containing only the plate constants as unknowns. In the case of a large photographic net covering a hemisphere or sphere with multiple overlap of the plates, the actual number of unknown plate constants may exceed 20000, depending on the adopted reduction model.

The general structure of a computer program, now under development at the Hamburg observatory, is described in detail. The program, entirely written in Fortran IV language, is based on a very fast subroutine, developed at the university of Stuttgart, for the solution and inversion of large symmetric and positive-definite linear systems of equations with banded-bordered coefficient matrix. The adopted solution procedure in a generalized Cholesky method and the programmed algorithm is based on a direct transfer of submatrices instead of single coefficients of the system of equations during the solution process. The program provides a direct non-iterative solution of the block-adjustment problem with at least partial inversion of the normal equation matrix for evaluation of the standard deviations of the computed star positions at selected block-points.

Block-adjustment methods will provide maximum advantage with respect to error propagation and required minimum number of reference stars, if the number of unknown parameters is small.

The choice of suitable functional and stochastic reduction models and the application of the program for theoretical accuracy studies using inversion and simulation methods are discussed. The main aim of such theoretical error studies is to investigate the dependence of the positional accuracy obtained from the block-adjustment on the number, accuracy and distribution of reference stars and the adopted reduction model. Estimates of computing effort and necessary external storage capacity for the

---

* Paper in press, *Monthly Notices Roy. Astron. Soc.* **167** (1974).

solution of the normal equations as a function of the number of unknowns and the actual matrix-structure are obtained.

For a variety of overlap patterns and block structures, occurring more frequently in photographic zone-work, simple analytic expressions for estimates of computing effort are derived. In addition, ordering schemes for an optimal arrangement of the single plates according to minimum bandwidth of the reduced normal equation matrix are provided.

Based on these results and adopting a CDC 7600 computer system, computing times and necessary external storage capacities are obtained for several large photographic nets, including a hemisphere and the full sphere with 2-fold and 4-fold overlap. In all cases a center-edge overlap pattern and a plate size of $5° \times 5°$ has been assumed. It is shown that a block-adjustment of the largest net, a full sphere with 4-fold overlap, could be provided within 3 hours computing time for the solution of the reduced normal equations containing 48 000 unknown plate constants.

LA REDUCTION DES CLICHES ASTROGRAPHIQUES DE CHAMP QUELCONQUE

J. DOMMANGET

Observatoire Royal de Belgique, Belgium

Résumé. La réduction des clichés astrographiques nécessite l'emploi de formules dont le choix relève du problème de la représentation mathématique de phénomènes expérimentaux.

Or, chaque fois que l'on a affaire à un phénomène expérimental, on est logiquement tenu de tenter de le représenter par des formules qui expriment aussi correctement que possible, les mécanismes qui en sont les composantes. En particulier, il est formellement contraire à ces principes, d'imposer à la représentation, un nombre de degrés de liberté différant du nombre de degrés existant dans le problème physique, sous peine d'obtenir une représentation pouvant être illusoire.

Il faut aussi se garder de donner à la formulation, des caractéristiques par trop différentes de la formulation réelle.

C'est pourquoi, il est indispensable d'introduire dans cette formulation, tout ce que l'on sait des aspects physiques du phénomène et de ne recourir à l'empirisme qu'en tout dernier ressort.

Or, dans le cas de la réduction des clichés astrographiques, une étude détaillée de la question montre que même pour des champs s'étendant théoriquement jusqu'à 90°, le nombre de paramètres indépendants est au maximum de onze: le coefficient a de la réfraction (voir notre exposé sur la réfraction); les deux coefficients de la distorsion; l'échelle du cliché; les coordonnées du point de percée de l'axe optique dans le plan du cliché; les deux paramètres d'orientation du plan du cliché par rapport au plan focal; l'angle d'orientation des axes de mesure et les coordonnées du zéro de la machine à mesurer.

Ce nombre de paramètres indépendants décroît évidemment avec l'importance du champ couvert par les clichés.

Il est donc inconcevable d'adopter pour la réduction de clichés couvrant des champs plus petits, des formules comportant pour l'ensemble des deux expressions de x et de y, parfois plus de 10 paramètres indépendants et dont la forme, en plus, n'est pas parfaitement justifiée.

Les formules générales de réduction sont données.

Le cas des clichés de la Carte du Ciel est discuté, à titre d'exemple.

Gliese, Murray, and Tucker, 'New Problems in Astrometry', 309. All Rights Reserved.
Copyright © 1974 by the IAU.

THE YERKES OBSERVATORY PHOTOELECTRIC
PARALLAX SCANNER

W. F. VAN ALTENA

Yerkes Observatory, Williams Bay, Wis., U.S.A.

Abstract. A photoelectric parallax scanner designed and constructed at the Yerkes Observatory to measure parallaxes and proper motions at the telescope is described. The concept of the parallax scanner is similar to Griffin's radial velocity spectrometer except that the parallax star and the field stars are monitored separately, thus making the technique purely differential. Observations with the 40-in. refractor under less than ideal conditions gave the error of a single measurement as $\pm 4.5\,\mu$ (s.e.). Observations under clear skies and with improved drive and guidance should decrease the error by a factor of 2 to 3.

A prototype instrument designed to measure parallaxes and proper motions photoelectrically at the telescope has been constructed at the Yerkes Observatory. The concept of the parallax scanner is similar to Griffin's (1967) radial velocity spectrometer except that the problem is two dimensional. In Griffin's spectrometer a photographic positive of a stellar spectrogram is placed in the focal plane and scanned to and fro to yield the position of maximum transmission, which can be related to the stellar radial velocity relative to standard stars observed on the same night. The Yerkes Observatory photoelectric parallax scanner, which has been briefly described earlier (van Altena, 1972), utilizes a high contrast contact positive of a recent photographic plate of the stellar field in question. The mask is completely opaque except for the star 'holes', thus minimizing the night sky light due to the Moon, etc.

In Figure 1, which is an optical schematic of the parallax scanner, light from the 40-in. refractor objective is focused on the mask; when the mask is in registration with the stellar field, light is transmitted by the mask and imaged by the 140 mm lucite field lens on two photomultipliers, which are located in a cold box. The central 12 mm of the lucite lens has been cored out and beveled on the flat side to deviate the light from the central (parallax) star to an off axis photomultiplier, while light from the field stars is imaged on the central photomultiplier. The signals from the two photomultipliers are amplified and then recorded on a two-pen chart recorder. Figure 2 shows the tailpiece of the 40-in. refractor with the parallax scanner mounted on the automatic guiding camera.

The lower scan shown in Figure 3 is of BD $+42°3123$, a star of magnitude $V=8.34$, while the upper scan is of the field. In order to retain a small image diameter on the mask, BD $+42°3123$ was reduced to $V=12.3$ by a 4-mag. sector on the mask plate. The scans shown in Figure 3 were made on a night with light cirrus, the effects of which can be seen in the tracings. Much of the noise in the tracings is due to telescope drive errors and 'hunting' in the automatic guider, about $\pm 4\,\mu$ rms, which is unimportant in the photographic work. We are making a number of improvements in the telescope drive and guider to minimize the hunting since it is deleterious to the parallax scanner observations. The scan rate used at the present time is a slow

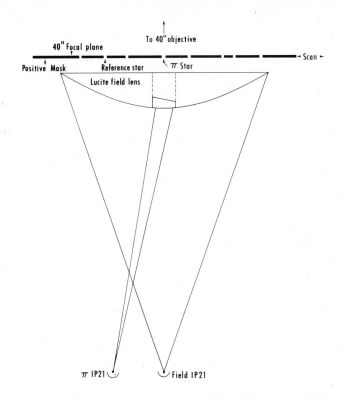

Fig. 1.   Optical schematic of the Yerkes Observatory parallax scanner. See text for explanation.

$1''$ min^{-1} of time in order to average out guiding errors and fluctuations in seeing.

Measurement of the position of the central star relative to the field is done by measuring the peak displacements on the chart paper and multiplying by the appropriate scale factor, which is 5.2 μ/chart mm in this case. Based on the interagreement from six scans of this star an error of $\sigma_1 = \pm 4.5$ μ (s.e.) for the relative position from a single scan was found. This is about two times poorer than we can do photographically at the present time at Yerkes, where $\sigma_1 = \pm 2.3$ μ (s.e.) is average. However, once the guiding errors have been minimized and given clear weather it should be possible to reduce the parallax scanner error to less than 2 μ for a single scan. Further improvements could undoubtedly be made through pulse counting and computer analysis.

Aside from the high potential accuracy of the technique, due to its purely differential nature, a very large gain has been achieved in reducing the effort involved in parallax determinations. A parallax series with the parallax scanner would typically consist of perhaps six scans on a night with perhaps 25–30 nights spread over 2–3 years as is presently the case for most Yerkes parallaxes. However, the tracings can be measured and reduced in a few minutes in contrast to the large amount of time

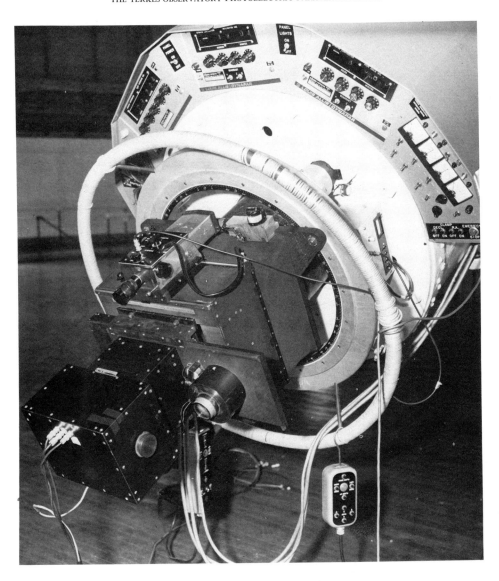

Fig. 2. The tailpiece of the 40-in. refractor showing the parallax scanner mounted on the automatic guiding camera.

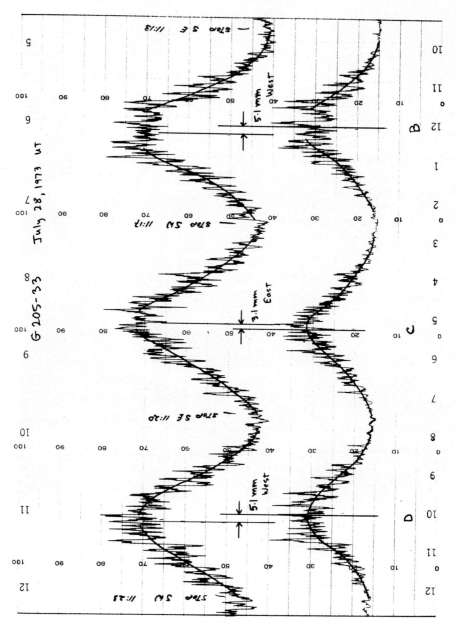

Fig. 3.   Three scans of the field of BD +42°3123 ($V=8.3$) with the parallax scanner. The scans are affected by noise introduced by light cirrus, guiding errors and seeing fluctuations. The error of a single relative position is $\pm 4.5\ \mu$ (s.e.).

spent measuring and reducing the plates; the reduction in time should be $\sim 90\%$.

Some disadvantages of the technique are the lack of a photographic plate to re-analyze perhaps 50 years later, no information on the individual reference star relative parallaxes and proper motions can be obtained, and it may not be possible to observe either close binaries or stars with proper motions larger than $\sim 0\rlap{.}''2$ yr^{-1}. Despite these limitations observations with the parallax scanner look promising and work will be continued towards developing a truly operational model.

Acknowledgements

I would like to acknowledge the major effort put into developing the parallax scanner by Robert J. Pernic.

This research has been supported in part by the National Science Foundation under grant NSF GP-13771.

References

Griffin, R. F.: 1967, *Astrophys. J.* **148**, 465.
Van Altena, W. F.: 1972, *Bull. Am. Astron. Soc.* **4**, 11. (Observatory reports.)

DISCUSSION

Moffet: Griffin's method for measuring radial velocities is useful because the same mask can be used for hundreds of stars, for example in a cluster. I fail to see the advantage of your system, which requires a special mask for each field. Is it in any way better than conventional plates and an automatic measuring engine?

Van Altena: Since it is necessary to make about 25 observations of a parallax star over a period of 2–3 years the requirement for a new mask for each field is not too burdensome. Advantages of the parallax scanner are that the observations are purely differential and not semi-absolute as in the radial velocity spectrometer, and that even with automatic measuring machines the amount of labor involved in the measuring and reductions to determine a high quality parallax is excessive. The parallax scanner can reduce this effort by about 90%.

ON INCONSISTENCIES FOUND IN LONG-TERM PARALLAX SERIES

W. D. HEINTZ

Dept. of Astronomy, Swarthmore College, U.S.A.

Abstract. Two recent papers (Hershey; Gatewood and Eichhorn; *Astron. J.*, 1973) pointed out that:

(1) A proper motion irregularity interpreted as orbital motion was found duplicated in another field.

(2) The discontinuity found at one telescope failed to be confirmed by others, and might be ascribed to mere instrumental errors.

After the duplication was discovered two years ago, further measurements have revealed an identical pattern in at least seven parallax stars. Yet the cause of the effect, and the reason for its absence in other cases, have not been located. Colour effects and filter problems appear to be ruled out. Indications for discontinuities occurring at other epochs were found. Thus, the question cannot yet be safely answered whether the small variations (one or two micron) of proper motions reported in some cases are spurious; in any case, great caution should be exercised in interpreting one-micron effects unless confirmation from another instrument is obtained.

Gliese, Murray, and Tucker, 'New Problems in Astrometry', 317. All Rights Reserved.

PRECISE REDUCTION TO THE APPARENT PLACES
OF STARS*

S. YUMI, K. HURUKAWA, and Th. HIRAYAMA

International Latitude Observatory, Mizusawa, and Tokyo Astronomical Observatory, Japan

Abstract. For a precise reduction to the apparent places of the stars in a uniform system during the 19th and 20th centuries, the 'Solar Coordinates 1800–2000' by Herget (*Astron. Papers* **14**, 1953) may conveniently be used, because no coordinates of the Sun, referred to the mean equinox of 1950.0, are given in the Astronomical Ephemeris before 1930.

A maximum difference of $0''.0003$ was found between the aberrations calculated from both the Astronomical Ephemeris and Herget's Tables for the period 1960–1969, taking into consideration the effect of the outer planets, which amounted to $0''.0109$.

The effect of the inner planets on the aberration is estimated to be of the order of $0''.0001$ at the most and the correction for the lunar term due to the change in astronomical constants is $0''.00002$. It is recommended that the solar coordinates be calculated directly from Newcomb's formulae taking the effects of all the planets into consideration, but the effect concerned with the Moon can be neglected.

DISCUSSION

Lederle: The discrepancies found in the solar coordinates are of the same order of magnitude as the precision limit of both Newcomb's expressions and tables. It is therefore questionable whether one reaches a real improvement by the proposed corrections.

Yumi: I accept your point, as otherwise anyone could expand the theory independently of Newcomb. We could, however, confirm in what way we can obtain the apparent places of the stars in a uniform system throughout a long period of time, including a period before 1930.

It should be noted, however, that the longitudes of Venus and Jupiter given in Newcomb's Table might have an error amounting to $0''.07$, the details about which will be written by H. Kinoshita of the Tokyo Astronomical Observatory in a forthcoming paper.

Lederle: The aberration vector (ΔR) has to be applied to a unit vector, and the expression on the right hand side of Equation (12) should therefore be multiplied by a rectification factor before ΔR is added, if full correctness is claimed.

Yumi: There is no problem because the calculation is actually made as you mentioned.

* The full text of this paper will appear in *Publications of the International Latitude Observatory, Mizusawa* **IX**.

GENERAL DISCUSSION,
SUMMARY AND RESOLUTIONS

GENERAL DISCUSSION

Elsmore: Further to the brief discussion we have already had, I wonder what advice you would give to the radio astronomers as to the best choice concerning the zero point of right ascension. Should radio astronomers strive to establish the equinox fundamentally, or is it sufficient to use an FK4–FK5 star or an extragalactic object, or is it better to adopt a mean from several stars or from several extragalactic objects?

Dieckvoss: I would suggest a few radio sources with optical counterparts.

Tucker: It may be preferable for the radio astrometrists to define their position zero in terms of a galactic rather than equatorial co-ordinate system. This would avoid the anomaly of setting up a precessing radio co-ordinate system when a non-precessing system is available, constituted by a source catalogue giving determined (l, b). Transformation from (l, b) to $(\alpha_{1950}, \delta_{1950})$ will then be by the rigorous definitions adopted by the IAU in 1959.

Dieckvoss: We should not forget that the radio measurements of R.A. differences and Declinations are made with antennae fixed to the rotating Earth.

Kovalevsky: The radio astronomers should build their own reference system independently of the classical definitions (reference to FK5) and observe links with stars and with the solar system as much as they can in order, when possible, to have observations permitting a link with the ecliptic and equator. But they should not try to define their system as optical people do (equinox, etc.) if we do not want to introduce difficulties and misunderstandings.

Eichhorn: Since right ascension and declination are basically determined by the kinematics of the Earth and completely extrinsic to the stellar system there is little merit in choosing this co-ordinate system exclusively as the basis for positions observed by radio means.

Gubbay: I think we may be approaching a philosophical limit in that as we approach position accuracies of $0''.001$ the objects we see are likely to suffer increasing variations in flux density and in position.

It seems advisable to adopt a group of reference sources which are extragalactic, are visible with VLBI at microwave frequencies and have optical counterparts pro tem, and to review this catalogue as experience grows in radio astrometry.

Wall: I want to raise a point in connection with radio astrometry. For several years at Parkes we have been carrying out a 2700 MHz survey for extragalactic radio sources, with the finding of compact sources as a principal objective. Several hundred such sources now appear in the Parkes catalogues, and some 300 have been identified with quasi-stellar objects. Thus in the sense that the sources for radio astrometry in the southern hemisphere are known, we are perhaps more fortunate than our northern hemisphere colleagues. However, the instrumentation for accurate measurement of the positions is lacking and the support of all members of this symposium is solicited in order to obtain it.

Matsunami: When we take into account the effects of tropospheric and iono-spheric scintillation and of diurnal variation of electron density of the ionosphere, we propose an interferometer which has a baseline of about 100 km and is to be operated with a frequency of about 5 or 6 GHz and has a phase lock system by two way radio link. Significance is just the same as Dr Elsmore's work. The problem is the cost of construction and of maintenance. If we could achieve this, we may get an accuracy of less than 1 ms (in time) for radio sources at $\delta = 0°$.

Van Herk: Statisticians complain that astrometrists are not handing down results in the way a statistician would like it: the compiler of a catalogue gives a result which is a modification of the original observations.

The statistician cannot apply proper statistical methods to this material. It is suggested that the individual observations, together with relevant auxiliary data – as far as we can know them – are kept available. Publishing individual data is of course, out of the question.

Eichhorn: It would give future investigators the full benefit of results of past statistical adjustments if not only the unknowns and their standard errors were published, but also the covariance matrix of these quantities. This would not require too much space and – from the standpoint of statistics – be virtually equivalent to communicating – in printing – all the original observations.

Klock: I can sympathise with Dr van Herk's request but in the case of a typical observing programme where an FK4 star may be observed 80–100 times, it would be impossible to publish particulars of each observation. Generally at the U.S. Naval Observatory we retain this more detailed data at least 10 years after the catalogue is published.

In recognition of their distinguished service to international projects in fundamental astrometry, telegrams of greeting were sent on behalf of the participants to Mr F. P. Scott, formerly of USNO, and Prof. M. S. Zverev, of Pulkovo.

SUMMARY AND CONCLUSIONS

BART J. BOK

Steward Observatory, Tucson, Ariz., U.S.A.

This has been a good Symposium. There obviously is a need to review now the problems of the accepted fundamental reference system of star positions and proper motions. The basic Fourth Fundamental Catalog (FK4) has been *the* reference catalog for the past 10 years. It needs updating and especially it should be made more directly usable for discussions of positions and motions referred to faint galaxies. In the preparation of the next catalog we should make use of radio galaxies as basic reference points for fixing precision stellar positions. The Symposium came at the right time! Radio Astrometry has burst upon the scene and it is essential that the optical and radio astrometrists should get to know each other and exchange views about the manner in which together we may work towards the establishment of a fundamental system of positions and proper motions more reliable than we have had in the past. There has also been much activity in the area of measuring proper motions of faint stars relative to galaxies and we obviously have to consider the best manner in which these valuable new contributions can be applied most usefully to basic astrometry. Our Symposium was held in the right place! It is high time that we should draw attention to the accomplishments and the future needs of basic southern hemisphere astrometry; Perth, Western Australia, is obviously a good place to discuss such matters. Finally we should discuss questions relating to instrumentation. Over the past decade there has been what could be called a 'breakthrough' in instrumentation, not only in the radio area, but also in the more traditional optical area of measurement by transit circles. There have been many new developments in the field of automatic measurement of photographic plates. This was clearly the time to take stock and plan for the future.

The Scientific Organizing Committee had sensibly picked seven areas for special discussion. I shall report briefly on each of these in order of the Perth presentations.

A. REFERENCE SYSTEMS

Ten years ago the Rechen-Institut in Heidelberg, of which the Chairman of the Scientific Organizing Committee, Walter Fricke, is the Director, published the Catalog FK4 with precision positions and proper motions for 1535 fundamental stars. FK4 needs updating and we must have, in addition, data of comparable quality, firmly tied into our basic fundamental system, for many additional stars and for some faint galaxies. Our present requirements are twofold. First, we need urgently a Supplement to FK4 that will include precision data on the FK4 Reference System for a few thousand faint stars. This Supplement should include precision positions

Gliese, Murray, and Tucker, 'New Problems in Astrometry', 325–333. All Rights Reserved.
Copyright © 1974 by the IAU.

on the FK4 System for a number of star-like galaxies as well. Second, we must get in shape for FK5, which will require that we obtain precision positions for the epoch of 1975, more or less, for most of the stars in FK4. Furthermore, we must obviously check and improve as well the quality of the proper motions for the stars in FK4.

We had much useful discussion during our Symposium about the need to determine afresh the basic values for the precessional corrections in use today. The adopted values for the precessional constants that we use today are still those derived by Simon Newcomb early in this century. I am sure that Newcomb would be pleased to learn, if he could be so informed, that his constants have continued to be the basic ones for about three-quarters of a century. But I think, too, that Newcomb would be the first one to suggest that the time has come for slightly different values to be adopted for use in compiling FK5! It seems from our discussion that the planetary precession is probably fixed with sufficient accuracy already, but that the luni-solar precession needs careful overhauling. Along with this, we must look into new determinations of the Oort constants of galactic rotation, A and B.

W. Fricke and W. Gliese indicated that 1980 is the present target date for the publication of FK5. They reported that 150 dependable new catalogs of position have been published since 1950, and these must obviously be incorporated in FK5. The present plan is to include in FK5 all observations that will become available before the end of 1975. The work on a Supplement to FK5 should proceed in parallel with the work on the basic Catalog. Our Symposium suggested that there is a need for a Supplement containing 3000 to 5000 stars. The new requirements presented by radio astrometry, and also those for positions and motions of faint galaxies, should receive special attention. Present systematic uncertainties in proper motions of reference stars are considerable. Errors of the order of $0\overset{''}{.}005$ to $0\overset{''}{.}007$ per year may well be present. However, with continuing effort and insistence on precision positions and proper motions, such errors may well be reduced to $\frac{1}{3}$ their present values for FK5. Proper motions of precision with small systematic and accidental uncertainties are absolutely essential for the calibration of the distance scale of our Milky Way System, for the basic calibration of the HR diagram and for all sorts of problems relating to the calibration of absolute magnitudes of the stars. The astrophysicist, the student of galactic structure and the astronomers involved in the study of distances of faint galaxies all require as a basis for their work a good fundamental system of reference positions and proper motions.

B. SOUTHERN HEMISPHERE REFERENCE SYSTEMS

Problems relating to positions and proper motions of stars south of declination $-35°$ received special attention at our Symposium. When FK4 was prepared, the situation with regard to positions and proper motions of southern stars was in a sad condition. The only observatory south of the equator that contributed effectively to FK4 was the Cape Observatory in South Africa. The principal reason why our Symposium is being held in Perth is that, by doing so, special attention would be drawn to southern hemisphere needs. Moreover, the astronomers gathered at remote Perth wish to express their thanks and congratulations to their colleagues who have worked so hard during the last decade to remedy the existing deficiencies in the south.

New southern observations figured prominently in our discussions at Perth. B. J. Harris, the Director of Bickley Observatory in Western Australia, and his colleagues, notably I. Nikoloff, have worked on the problem in close association with a group of astronomers from Hamburg, Germany. In the later 1960's our German colleagues brought a precision transit circle to Western Australia for work on the southern heavens. We learned that 110 000 observations have been obtained with this instrument and that all of these have been reduced and analyzed. In addition, we now possess basic data from 45 000 absolute positions determined by the Chilean and Soviet astronomers jointly working at Cerro Calán Observatory near Santiago. The results of this magnificent work were reported for the two groups by the Santiago Director, C. A. Anguita. And, third, the U.S. Naval Observatory has completed the making of 115 000 observations of southern stars with a transit circle at the El Leoncito Station in Argentina. The various groups working from the southern hemisphere are agreed that large corrections $\Delta\alpha_\delta$ must be made to FK4 positions in the South Polar Cap.

We cannot expect to rest on our laurels. We now possess a good set of basic positions for southern reference stars for the epoch 1970, but, to obtain good proper motions for these stars, a continuing major effort is obviously called for. If we diminish our southern hemisphere efforts at this time, then we are asking for trouble by the year 2000, for we must obviously have for all of our reference stars good proper motions along with high quality positions. We must continue observational positional work from several southern hemisphere observatories. We were delighted to learn of the excellent progress being made in the preparation of an extensive catalog of Southern Reference Stars.

C. RADIO ASTROMETRY

Our Symposium will be remembered principally as having been the occasion for the first meeting of radio and optical astrometrists. One of our principal tasks has been to consider how the results of radio astrometry can be incorporated in our researches on fundamental systems of reference for positions and proper motion. Among our colleagues attending the Perth Symposium were about a dozen radio astronomers. Three of them presented major review papers: C. M. Wade of the National Radio Astronomy Observatory of Charlottesville, Virginia, and Greenbank, West Virginia, B. Elsmore of the Mullard Radio Astronomy Observatory of Cambridge, England, and C. C. Counselman of Massachusetts Institute of Technology. The three took the time to explain in detail to the optical astronomers the techniques of observation used by radio astrometrists, a gesture that was both effective and much appreciated by all present.

In his Introductory Remarks, Chairman Fricke noted that a year ago the methods of determining absolute radio positions were near the point where they could compete with the methods of traditional fundamental astrometry. He said further "I find that this point has already been reached and that the radio methods have been developed to such a perfection that an even higher accuracy can be reached than in optical absolute measurements." At the conclusion of our Symposium I feel free to report that all of us present at Perth wholeheartedly agree with Fricke's evaluation.

The most impressive results to date have been obtained with the aid of connected element radio interferometers involving moderate baselines of a few kilometers. Absolute values for the declinations of the radio sources can be found from the measurements of the rates at which these sources move through the interferometer patterns. What the radio astronomer does basically when he fixes a declination, is to determine a very precise radius for the daily circle described by the radio source under investigation. He requires no optical measurements of position for reference purposes. Absolute declinations come automatically from his observational data. Precise time measurement is fundamental to success in radio astrometry. The radio astronomer can measure large arcs in the heavens with the same precision as small arcs, and differences in right ascension can therefore be very precisely observed. The radio astronomer cannot observe the Sun and planets with the same precision as that obtainable for radio stars and galaxies, and asteroids are not suitable (at least not yet!) as basic reference points. The radio astronomer is therefore not capable yet of fixing directly the zero point of his system of right ascensions, the vernal equinox, and he must turn to his optical colleagues for zero point determinations. This requires that special emphasis must be given to the determination of precision optical positions for radio sources that have been identified optically. Such sources are (1) all radio stars, (2) radio galaxies and (3) quasars.

Highly dependable positions for radio sources have been obtained. Presently determined radio positions have uncertainties in right ascension and in declination which are in the range of $\pm 0\rlap{.}''1$ to $\pm 0\rlap{.}''01$. Data of such precision are now available for close to 100 radio sources, spread evenly over the sky from the North Pole to about 40° southern declination. Three good radio stars, Algol, Beta Lyrae, and P Cygni are already known and more are promised; their fundamental optical positions in the reference system of FK4 (and soon of FK5) must obviously be measured with high precision. We must not overlook the pulsar in the Crab Nebula as an optically identified radio source! The optical astrometrist has as one of his major future assignments to determine with minimum delay a good fundamental position for any star that is identified as emitting radio radiation. Many optically identified compact galaxies emit radio radiation, and the optical positions for these objects must of course become available also with minimum delay. The same applies to the quasars which are shown on our photographs and which often emit strong radio radiation.

While the work in radio astrometry for moderate baselines is already yielding first-class results, even greater precision is ultimately promised from observation using the techniques of Very Long Baseline Interferometers, some of them intercontinental. At Perth, Counselman indicated that he is of the opinion that ultimately precisions of the order of $\pm 0\rlap{.}''005$, or better, may be obtainable. But this is still in the future. It became clear during the discussion at Perth that, for the effective use of radio positions already measured or about to become available, optical astrometrists must provide through the Supplements to FK4 and FK5 high caliber positions not only for *all* radio stars and for most optically identified radio galaxies, but that there is furthermore a need of a good many good standards between 16th and 19th apparent magnitude for stars near radio galaxies.

Much interest was expressed at the Perth Symposium in the results of the University of Texas radio astrometry survey, reported by J. N. Douglas. With the aid of the Texas 120-channel radio interferometer, source positions have already been determined to within one second of arc for 2000 radio sources and 50000 more such sources are within reach of this equipment with comparable precision.

Once again the southern hemisphere is being discriminated against. The radio interferometers that have been most effective to date have all been in the northern hemisphere. The Perth Symposium passed unanimously a resolution urging the extension of the application of radio astrometric techniques to southern hemisphere observations. If this is not done soon, then we shall have to live for the next twenty years or so with a lopsided fundamental system of reference for positions and proper motions. The northern hemisphere will be in good shape and the south will be lagging behind. All astronomy and astrophysics will suffer as a consequence.

One of the great things that developed from the meeting at Perth was that it brought together optical and radio astrometrists – who realize that one cannot succeed without the other.

D. ASTROMETRY WITH LARGE TELESCOPES

E. PROPER MOTIONS AND GALACTIC PROBLEMS

During our Symposium, many important papers were presented in the sessions under the above headings. We devoted a full day of our Symposium to the problems of astrometry with large telescopes, especially the measurement of proper motions and trigonometric parallaxes of faint stars, and the use of this material in Milky Way research.

In a way, the most significant presentations related to the measurement of positions and proper motions with reference to galaxies. The vast complex of researches initiated by W. H. Wright at Lick Observatory in California, and by A. N. Deutsch of Pulkovo Observatory in the U.S.S.R. have produced already some very important and impressive results. A. R. Klemola and Deutsch reported on the Lick and Pulkovo programs. The Lick Observatory Program is well on the way toward completion. Most first and second epoch photographs have been made with the twin 20-in. photographic and photovisual refractors at Lick Observatory. The separation between first and second epoch plates is mostly of the order of 20 years. The Lick method consists of measuring relative positions of a considerable number of stars faint galaxies on a pair of plates, and from these positions the proper motions of the stars can then be obtained with reference to the galaxies. Close to 60000 stars and faint galaxies are being measured at Lick, 30000 of them stars of "special interest" (including 2400 RR Lyrae stars) and 29000 stars that possess good reference positions in the system of FK4, mostly from AGK3. These will ultimately be related directly to the positions in the system of FK5. At Pulkovo Observatory, Deutsch and his associates have completed measurements of positions of stars relative to star-like knots in 85 spiral and other external galaxies.

The first result of intercomparisons between Lick proper motions and the Pulkovo proper motions were reported at our Symposium. These results are not cheerful. There seem to be considerable systematic differences between the annual proper motions determined by the two rather different methods, with occasional systematic differences in R.A. being found as great as $0\overset{''}{.}010$ yr^{-1}. This is a matter that clearly requires prompt further examination, and we are assured that such examination will be forthcoming soon. The techniques developed and applied at Lick and at Pulkovo Observatories obviously have terrific importance, for the results obtained by these methods should lead to a firm tying-in between positions and proper motions for the brighter stars and for faint stars and galaxies. In the preparation of the Supplements to FK4 and FK5 careful attention will have to be given to the inclusion of a good number of Lick and Pulkovo stars (and galaxies) on these lists.

During the session on proper motions and galactic problems, S. V. M. Clube of Edinburgh presented a paper that shows that very new and unexpected results may come from analyses of proper motions of faint stars observed in the Lick survey. He seems to have detected unexpected local streaming tendencies within 300 pc of the Sun on the basis of his analysis of the first list of 8000 Lick proper motions.

We heard during our Symposium a good deal about three related projects. Anguita announced that a survey in principle not unlike the one at Lick Observatory is being undertaken at the U.S.S.R. Station at Cerro El Roble in Chile (with a Maksutov Telescope) and a second northern survey is planned with the telescope of Tautenburg Observatory in the German Democratic Republic. All of this was very welcome news, but not so welcome was the information that the Yale-Columbia Project at the El Leoncito Station in Argentina is in danger of being prematurely stopped by lack of funds. The Yale-Columbia effort, which to date has been supported by the National Science Foundation, may not even have the funds for the completion of its whole first epoch program of photographic and photovisual plates. In a general outline, the Yale-Columbia program follows closely the precepts of the Lick Program and ultimately we may expect from it for the southern hemisphere results of a scope and precision comparable to those now becoming available from the Lick Program. But to do the job right, we need now to see to it that the first epoch plates are completed without delay and that these plates should be of the required high quality. The initiation of the program for the taking of second epoch plates can be postponed for a while, for at least 15 years should elapse between the taking of first and second epoch plates for proper motion work. However, it would be a disaster if the first epoch program would not be fully completed now. The Symposium at Perth passed a resolution showing its deep concern regarding the possibility that another southern hemisphere project may be neglected. A resolution in support of the completion of the Yale-Columbia Program was also passed by IAU Commission 24 during the

General Assembly in Sydney, which was held during the two weeks after the Perth Symposium.

W. J. Luyten received an ovation for the magnificient work that he has done on locating and studying faint stars of large proper motion with his new fully automatized and computerized plate scanner and measuring engine. He offered a list of 11 000 stars with proper motion in excess of $0\overset{''}{.}18$ yr^{-1}, and another list of 800 new stars with proper motions in excess of $0\overset{''}{.}5$ yr^{-1}. He noted that his latest maximum of the General Luminosity Function is at photographic magnitude $+15.4$ which is a little brighter than his earlier result, which placed the maximum at $+15.7$.

During our discussions frequent reference was made to the astrometric needs for access to large reflectors and Schmidt telescopes. Astrometrists need this access because it is only through the use of these instruments that such projects as the determination of faint membership of open and globular star clusters can be resolved. Unless such access is provided now, the first epoch plates for proper motion work will not be available 20 years hence when questions will be asked about such faint membership. In a way the most satisfactory solution is that achieved by the U.S. Naval Observatory. At its Flagstaff Station it operates the 61-in. astrometric reflector, a truly magnificent instrument for work of this nature. Harley Wood, the Government Astronomer for New South Wales in Australia, pointed out that a similar instrument is very much needed for astrometric work in the southern hemisphere. The 98-in. Isaac Newton Telescope at Herstmonceux is already being used 3 or 4 nights per month for astrometric research. C. A. Murray reported first results for parallax and positional work based on photographs made with this telescope. Very encouraging results were also reported by W. F. van Altena with the good old Yerkes 40-in. refractor.

We listened to various reports about the current status of research on trigonometric parallaxes. K. Aa. Strand reported that good progress is being made with the program of trigonometric parallaxes under way at the U.S. Naval Observatory's Flagstaff Station. 209 trigonometric parallaxes obtained from photographs made with the 61-in. astrometric reflector have already been published. The selection of future stars to be placed on the Flagstaff and other programs was discussed at length. Work on trigonometric parallaxes with large reflectors is so time-consuming that great care must be taken in advance in the selection of candidates for the measurement of trigonometric parallaxes. Large proper motion is a fine first criterion for selection, but every star chosen tentatively for such work should be examined carefully. If the star is bright enough, Schmidt reflector objective prism photographs should be examined or higher dispersion spectra should be obtained with image tube spectrographs. The Hoag-Schroeder technique, in which a transmission grating is placed close to the Cassegrain focus of a large reflector, promises to be very useful as well. Furthermore, every candidate should be subjected in advance to careful photo-electric photometry, which, for the redder stars, had best be done in visual and near infrared light.

During our discussions much attention was paid to the cutting down of the accidental errors in measured trigonometric parallaxes. In spite of much effort, most finished trigonometric parallaxes are uncertain by $\pm 0''.004$. Some further reduction in the mean errors obtainable from observatories on Earth may be achieved, but not much. However, I should report that space research holds great promise in this area. The Large Space Telescope should yield trigonometric parallaxes with errors on the order of $\frac{1}{3}$ of those now obtainable from Earth. Truly spectacular results may be expected when space missions with astrometric equipment extend the baseline through observations from the vicinity of Jupiter and simultaneously by observation from a distance of 5 AU in the opposite direction! The sample of our Milky Way system for which trigonometric parallaxes figure importantly at the present time lies entirely within 25 pc distance from the Sun. With much effort and proper care it now seems within attainable limits that our precision might go up sufficiently to double the radius of the sample, which would yield reliable data to distances of 50 pc from the Sun. With another doubling of precision through observation from space vehicles near the Earth, we should be able to move out to radius of 100 pc from the Sun, which would increase the volume of our space sample by a factor of 64 over that within reach to 25 pc distance. By extending our baseline to 5 AU, this volume could be increased by another factor 125, and our sample would be able to give us reliable trigonometric parallaxes for stars to 500 pc from the Sun.

F. ASTRONOMICAL REFRACTION PROBLEMS

G. ASTROMETRIC TECHNIQUES

During the final days of our Symposium, we had much discussion about improvements in techniques of observation for astrometric purposes, and we listened to several descriptions of new and ingenious measuring engines for obtaining photographic positions. The basic instrument for the measurement of stellar positions is still the meridian or transit circle. Its future was examined with care a year ago at an International Colloquium held at Copenhagen. G. van Herk reported about the discussions at Copenhagen, and it is helpful that his report will become part of our Proceedings. Whereas, early observers were proud to measure with transit circles stars of 10th apparent magnitude by simple visual techniques and recordings, it will soon become possible to obtain comparable accuracy to 12th mag. with the same instruments that have been used in the past. All of this is achieved through the application of photoelectric techniques and of automization of the recordings, a field of research and development to which the President of the IAU, Bengt Strömgren, has made some important and fundamental contributions. Modifications of the traditional transit instruments are being considered and have been developed by several astrometric groups. The horizontal mirror transits especially deserve attention. Our French colleagues spoke again very warmly about future uses of the Danjon astrolabe. Not only will this instrument be one of the basic pieces of equipment for use at the new astrometric observatory in the south of France (near Grasse), the 'Centre d'Etudes et de Recherches Géodynamique et Astronomiques' (CERGA), but we learned during our discussion that such astrolabes are being groomed especially for research in the neglected southern hemisphere. The instrumental development undertaken at the U.S. Naval Observatory, and at the astrometric observatories in Copenhagen, Hamburg, Ottawa and Pulkovo, deserve special attention for the future.

We have already referred above to Luyten's equipment for measuring and analyzing photographic plates. We listened to reports about several high speed automatic measuring instruments. The GALAXY Machine at Edinburgh, and other similar machines in Great Britain, promise to be exceedingly useful instruments for positional work. The automatic measuring machine, which has been in operation at the U.S. Naval Observatory for the past seven years, has proved its worth and a new version, incorporating laser interferometry techniques, is expected to go into operation soon. Lick Observatory has obtained a completely new version of its automatic measuring engine, which will soon be the instrument to measure the galaxy reference plates for the Lick Program.

H. CONCLUSION

It behoves a reporter of an IAU Symposium to end his remarks by presenting the principal conclusions that resulted from the discussions. On the last day of our meetings we took the time to discuss, occasionally rather vehemently, three major resolutions that we wished to bring to the attention to the whole of the astronomical world. These Resolutions represent the essence of our recommendations, and they show more clearly than anything that I can say or write, how effective and constructive a Symposium all of us have attended!

RESOLUTIONS

(1) This Symposium recognizes the inadequacy of the existing optical astrometric data to meet modern requirements for precision positions and proper motions. It therefore urges that the great accuracy of radio positions and the potentiality of new optical and space techniques should be fully exploited. Both absolute and differential observations are necessary, and it is particularly important that observations in the southern hemisphere should not be allowed to lag behind those in the north as they do at present in both optical and radio astrometry.

(2) This Symposium urges as the first priority the compilation by 1980 of an FK5 catalogue. The new reference frame must be extended to faint stars and extragalactic objects. This requires the vigorous continuation and, in fact, completion, especially in the southern hemisphere, of current programmes of photographic astrometry and, in addition, the development and exploitation of new photographic, radio and space techniques.

(3) This Symposium notes the existence of the Working Party established by Commission 40 to draw up a list of suitable objects for positional calibration of radio observations and urges co-operation between it and Commissions 8 and 24 to ensure as complete an integration as possible of the radio and optical astrometric fundamental programmes.